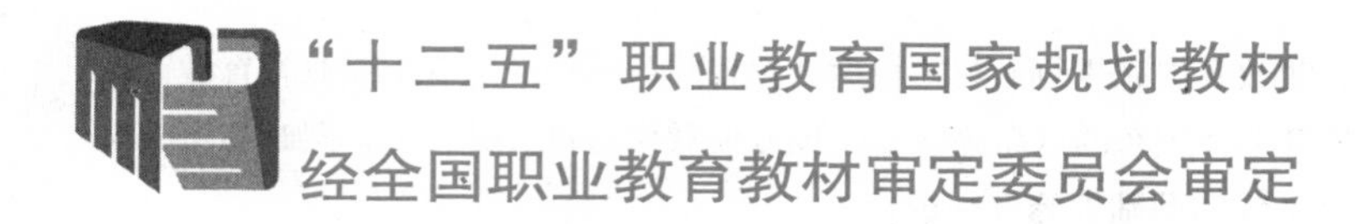

路由型与交换型互联网基础实训手册

第2版

主　编　程庆梅
副主编　徐雪鹏
参　编　陈　戌　王凯旋　徐　鹏　赵　鹏
　　　　陈中举　张　鹏　吴　丹

机 械 工 业 出 版 社

本书是“十二五”职业教育国家规划教材，根据《教育部关于“十二五”职业教育教材建设的若干意见》及教育部新颁布的《高等职业学校专业教学标准（试行）》，同时参考相关职业资格标准，在第1版的基础上修订而成。

本书主要分为2个部分，并细化为9章，内容以学生能够完成中小企业交换及路由网络调试实施和故障排除为目标。第1部分为第1～4章，主要介绍交换型网络功能实现以及功能强化；第2部分为第5～9章，主要介绍路由型网络功能实现以及强化。

为便于教学，本书配有电子课件，选择本书作为教材的教师可来电（010-88379194）索取，或登录网站 www.cmpedu.com，注册后免费下载。

本书可作为高等职业院校计算机应用专业和网络技术应用专业的实训教材，也可作为交换机、路由器管理和网络维护的配置指导书，还可作为计算机网络工程技术岗位培训的实训教材。

图书在版编目（CIP）数据

路由型与交换型互联网基础实训手册/程庆梅主编．—2版．—北京：机械工业出版社，2014.11（2015.8重印）

“十二五”职业教育国家规划教材

ISBN 978-7-111-48067-9

Ⅰ.①路…　Ⅱ.①程…　Ⅲ.①互联网络—路由选择—高等职业教育—教学参考资料　Ⅳ.①TN915.05

中国版本图书馆CIP数据核字（2014）第219729号

机械工业出版社（北京市百万庄大街22号　邮政编码100037）
策划编辑：梁　伟　责任编辑：蔡　岩
责任校对：张　力　封面设计：鞠　杨
责任印制：乔　宇
北京机工印刷厂印刷（三河市南杨庄国丰装订厂装订）
2015年8月第2版第2次印刷
184mm×260mm·11印张·262千字
3 001—6 000册
标准书号：ISBN 978-7-111-48067-9
定价：27.00元

第2版前言

本书是按照教育部《关于开展“十二五”职业教育国家规划教材选题立项工作的通知》，经过出版社初评、申报，由教育部专家组评审确定的“十二五”职业教育国家规划教材，根据《教育部关于“十二五”职业教育教材建设的若干意见》及教育部新颁布的《高等职业学校专业教学标准（试行）》，同时参考相关职业资格标准，在第1版的基础上修订而成的。

本书主要介绍交换型网络和路由型网络实现过程。其中第1部分介绍的是交换型网络中，针对不同功能需求交换设备的配置实施过程；第2部分介绍的是路由型网络中，针对不同功能需求路由设备的配置实施过程。本书在编写过程中力求体现工学结合，实训设计源于目前实际工程项目的真实需求；实训内容循序渐进、由简入繁；配置能力与排错能力并重的特色。本教材编写模式新颖，采用模块化设计，结构清晰，实训单元采用情景化设计，引发读者深入思考。

本书在内容处理上主要有以下几点说明：

1）教学实训环节中建议每章节2～4课时，并以小组为单位实施实训。全书建议课时68学时，授课讲师也可根据实际情况酌情安排。

2）教学实施中可以预留时间，组织学生进行内部讨论或实训总结。

3）由于本书以实训内容为主，建议与《路由型与交换型互联网基础》教材配合使用。

本书由程庆梅任主编，徐雪鹏任副主编。参与编写的还有陈戌、王凯旋、徐鹏、赵鹏、陈中举、张鹏、吴丹。

本书编写过程中，参阅了国内外出版的有关教材和资料，得到了北京市供销学校赵鹏、石家庄市职教中心黄琨老师的有益指导，在此一并表示衷心感谢！

由于编者水平有限，书中不妥之处在所难免，恳请读者批评指正。

编　者

第1版前言

伴随计算机网络的发展，各个行业都在积极地发展和升级计算机网络系统，通过提升办公网络的效率提高网上办公的效率，这也就需要越来越多的计算机网络工程技术人才投身到这个行业中，本书就是应这样的时代需求而编写的工程师入门级教材的配套实训手册。

- 指导思想

本实训手册特点在于将企业网络解决方案的产品融合在相对独立的实训过程中，每个实训又紧密地围绕着某个常见的网络需求，从而营造出一个与真实网络极其相似的网络搭建环境，读者通过仔细研读并跟随实训手册的步骤完成每个实训后，将体会到自身技术实力的提升。

- 本书的特点

1）注重实践操作，知识点围绕操作过程按需介绍。

2）每个实训均配套以情境介绍展开。

3）侧重实际能力的培养，抛开复杂的理论说教，学以致用。

- 编写思路

本实训手册共分以下几个部分：交换机实训，路由器实训，无线网络实训，安全相关实训，企业网络综合实训。每个部分均按照由浅入深的方式从简单的实训着手最终实现综合案例的分析和方案实施。

- 本书的读者

1）从事网络工程技术工作的初级网络工程师。

2）为终端客户提供网络搭建解决方案的网络工程师。

3）提供网络整体解决方案的售前售后工程师。

4）高等或中等职业技术院校的计算机相关专业二年级学生。

本教材全体编者衷心感谢提供各类资料及项目素材的神州数码网络工程师、产品经理及技术部的同仁，同时也要感谢来自职业教育战线的合作教师们提供的大量需求建议及参与的部分内容的校对和整理。

- 关于图标

本书图标采用神州数码图标库标准图标进行，除真实设备外，所有逻辑示意均使用如下图标。

高端路由交换机

机架式三层交换机

千兆三层交换机

千兆二层交换机

百兆三层交换机

百兆二层交换机

POE千兆交换机

通用网管交换机

核心路由器

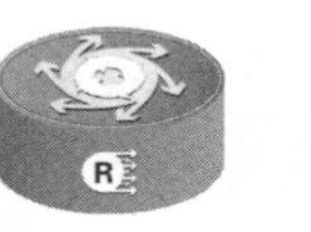

汇聚路由器

接入路由器

通用路由器

多核安全网关

WEB应用安全防火墙

通用防火墙

盒式AC

无线发射器

室外AP

机架式服务器

塔式服务器

笔记本

台式机

手机

限于编者的经验和水平，敬请使用本书的师生和各位同仁，对书中内容和文字上的种种缺陷和错误，提出批评。编者邮箱：dcnu_2007@163.com。

编　者

目 录

第 1 部分

交换机实训

第1章　交换机实践基础

实训1　交换机认识与带外管理

实训目标

小张中职毕业后在朋友开的网吧里当网管，后经熟人推荐来到某公司做网络中心的网络管理员。由于之前没有使用过网管型交换机，所以他的师父李工程师给他交代了以下3个任务目标：

1）熟悉网管型二层交换机的外观。

2）了解交换机各端口的名称和作用。

3）了解交换机最基本的管理方式——带外管理的方法。

实训拓扑

实训拓扑图如图1-1所示。

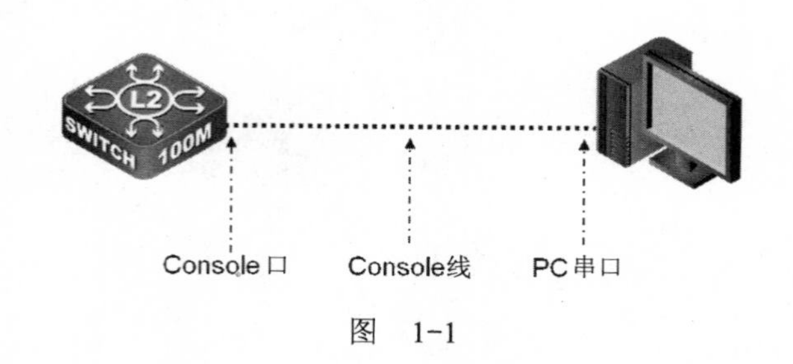

图　1-1

实训任务

任务1：观察交换机外观，认识交换机Console口及各网口的编号规则，如图1-2所示。

小张仔细观察了这种网管型交换机，通过向师父李工咨询，了解了交换机各端口的作用，知道了Console口就是用于带外管理的配置口，可连接在PC的串行口上对交换机进行管理。同时，知道了以太网口通常用三位数字表示：0/0/1中的第一个0表示堆叠中的第一台交换机，如果是1，就表示第2台交换机；第2个0表示交换机上的第1个模块，而第3

位数字则表示端口号。

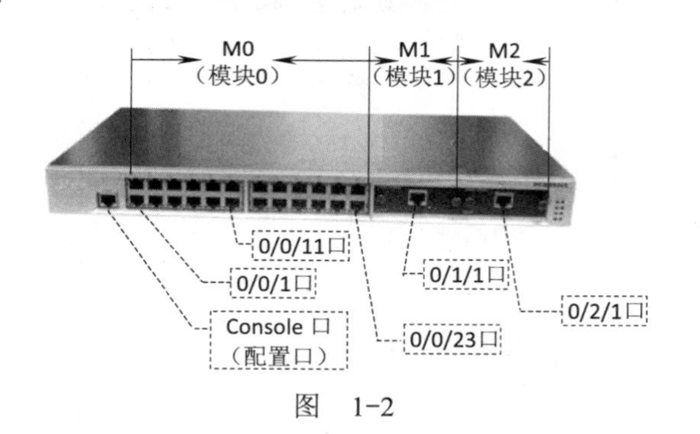

图　1-2

例如，0/0/1 表示的是堆叠中第 1 台交换机第 1 个网络端口模块上的第一个网络端口。默认情况下，如果不存在堆叠，交换机总会认为自己是第 0 台交换机。

提示：交换机的端口命名有 2 段式（x/x）和 3 段式（x/x/x）两种，具体的端口命名格式可以通过登录设备后使用 show running-config 命令查看。

任务 2：在关闭电源的情况下用 Console 线连接交换机与 PC。

小张观察了 Console 线，发现一端是 RJ-45 水晶头，另一端则是 RS-232 串行接口，于是他将 RJ-45 水晶头插入交换机的 console 口，另一端插入 PC 的 RS-232 串口。

任务 3：使用超级终端接入交换机管理界面。

小张打开了交换机与 PC 的电源，通过咨询师父李工，了解了接入方法。PC 启动完成后，单击“开始”→“程序”→“附件”→“通讯”→“超级终端”命令。

在打开的如图 1-3 所示的对话框中，小张为建立的超级终端连接取了个名字，李工告诉他，系统会把这个连接保存在附件的通讯栏中，以便于下次使用。单击“确定”按钮后，系统弹出“连接到”对话框，如图 1-4 所示。选择连接时使用的串口。

图　1-3

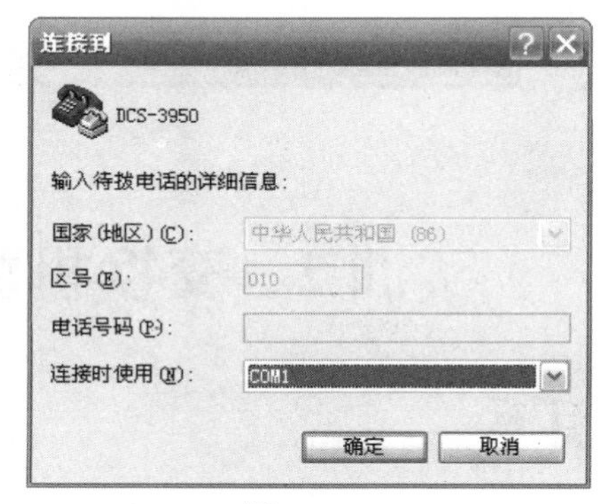

图　1-4

单击“确定”按钮后，弹出“COM1　属性”对话框，如图 1-5 所示。

通过咨询师父李工，小张在“每秒位数”下拉列表中选择了 9600，“数据位”选 8，“奇偶校验”选“无”，“停止位”选 1，“数据流控制”选“无”，单击“确定”按钮。

这时，就打开了超级终端连接窗口，按<Enter>键，进入如图 1-6 所示的交换机 CLI。

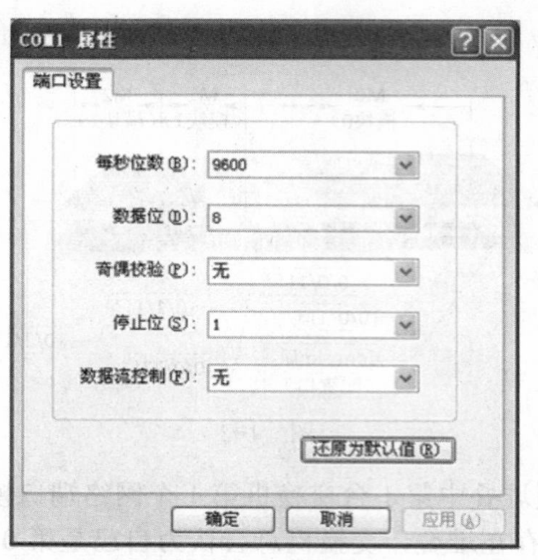

图 1-5

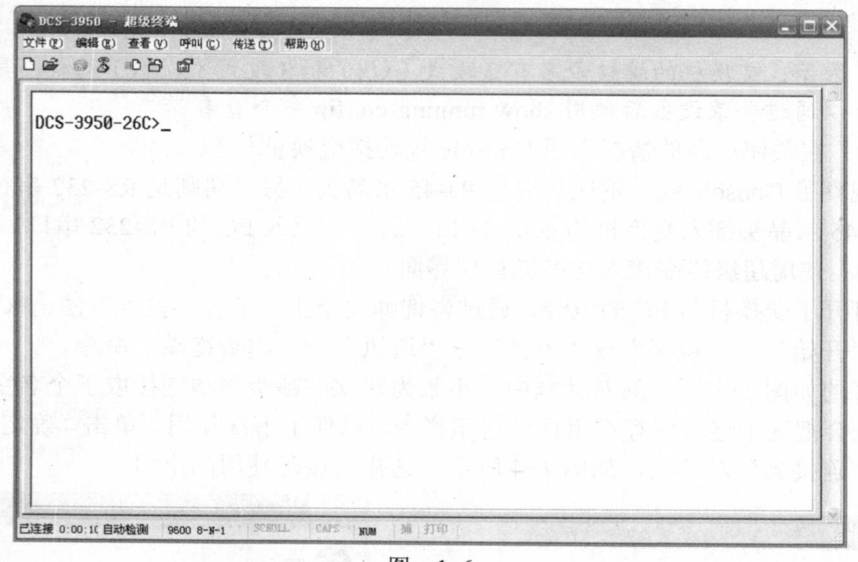

图 1-6

实训 2 交换机配置模式与 CLI 调试

实训目标

李工告诉小张，CLI（Command Line Interface，命令行界面）和 GUI（图形界面）相对应。CLI 由 Shell 程序提供。CLI 由一系列的配置命令组成的。根据这些命令在配置管理交换机时所起的作用不同，Shell 将这些命令分类，不同类别的命令对应着不同的配置模式。

CLI 是交换机调试界面中的主流界面，基本上所有的网络设备都支持命令行界面。国内外主流的网络设备供应商使用很相近的命令行界面，方便用户调试不同厂商的设备。神州数码网络产品的调试界面兼容国内外主流厂商的界面，和思科命令行接近，便于我们学习。只有少部分厂商使用自己独有的配置命令。

小张，我们来熟悉一下各种配置模式和 CLI 的调试。

实训拓扑

实训拓扑图如图 1-7 所示。

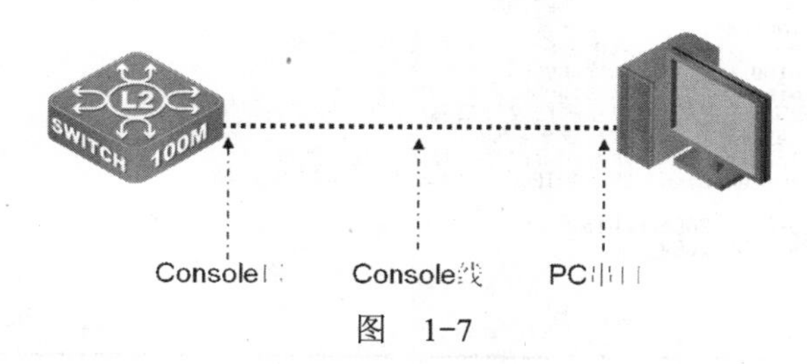

图 1-7

实训任务

任务 1：一般用户配置模式。

观察实训 1 出现的 CLI，发现提示符由两部分组成，前面是交换机名称，这个名称可以改变；后面跟着一个大于号“>”，这个“>”号表示当前交换机处于一般用户配置模式。之所以要把它称为一般用户配置模式，是因为任何人通过 Console 口接入交换机都可以进入到这个模式。

在一般用户配置模式下，可用的命令比较少，输入“？”后，即可看到该模式下可用的所有命令，如图 1-8 所示。

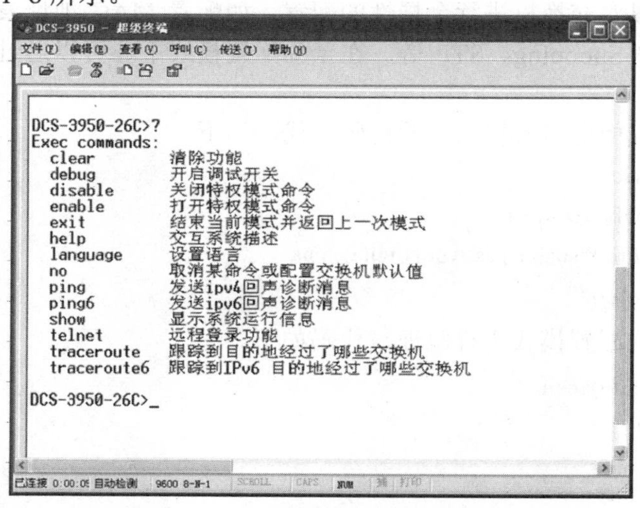

图 1-8

任务 2：特权用户配置模式。

在一般用户配置模式下，输入命令“enable”并按<Enter>键，可以看到提示符变成了

“#”，如图 1-9 所示。

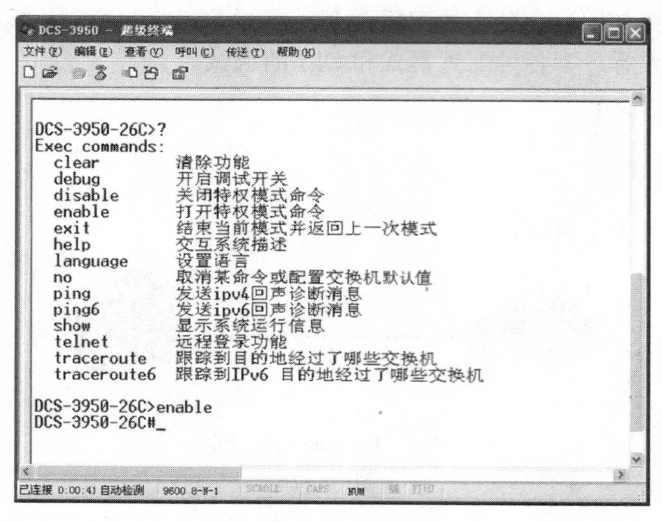

图 1-9

提示符“#”表示当前交换机处于特权用户配置模式。在特权用户配置模式下，可以查询交换机配置信息、各个端口的连接情况、收发数据统计等。而且进入特权用户配置模式后，可以进入到全局模式对交换机的各项配置进行修改，因此进行特权用户配置模式必须要设置特权用户密码，防止非特权用户的非法使用，对交换机配置进行恶意修改，造成不必要的损失。

要从特权配置模式返回到一般用户配置模式，可以使用“exit”命令。

任务 3：全局配置模式。

在特权用户配置模式下，使用“config terminal”命令可以进入到全局配置模式。在全局配置模式中，可以对交换机进行全局性的配置，如配置 MAC 地址表、端口镜像、创建 VLAN、启动 IGMP Snooping、STP 等。在全局配置模式下还可以通过命令进入到端口对各个端口进行配置。

在全局配置模式中配置特权用户的密码，代码如下。

```
DCS-3950-26C>enable
DCS-3950-26C#config terminal
DCS-3950-26C(config)#enable password digitalchina
DCS-3950-26C(config)#
```

返回到一般用户配置模式进行验证，代码如下。

```
DCS-3950-26C(config)#exit
DCS-3950-26C#exit
DCS-3950-26C>enable
Password:
DCS-3950-26C#config terminal
DCS-3950-26C(config)#
```

第一个 exit 是从全局配置模式返回到特权用户配置模式，第二个 exit 是从特权用户配置模式返回到一般用户配置模式。然后输入命令 enable，出现密码提示，输入刚才创建的特权密码 digitalchina 即可进入特权用户配置模式。再输入 config terminal 便进入了全局配置模式，如图 1-10 所示。

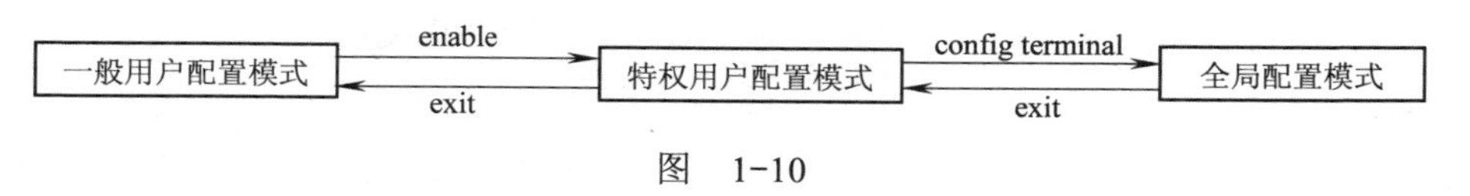

图　1-10

“师父，这三种模式我懂了，可这些命令太难记了！”

“初次接触都会有这种感觉，但 CLI 有捷径可走，根本不用记忆。下面我给你介绍一些 CLI 的技巧。”

任务 4：熟悉“？”的使用。

前面熟悉了三种配置模式，在全局配置模式中还有端口模式、Vlan 模式等子模式，这些在后面的任务中再介绍。现在，先来熟悉一下“？”符号。

在实训 1 中已经接触过了“？”，在“>”提示符下输入“？”便列出该模式下的所有命令。这个方法可以用在任何模式中。例如，在特权用户配置模式下输入“？”，屏幕显示如图 1-11 所示。

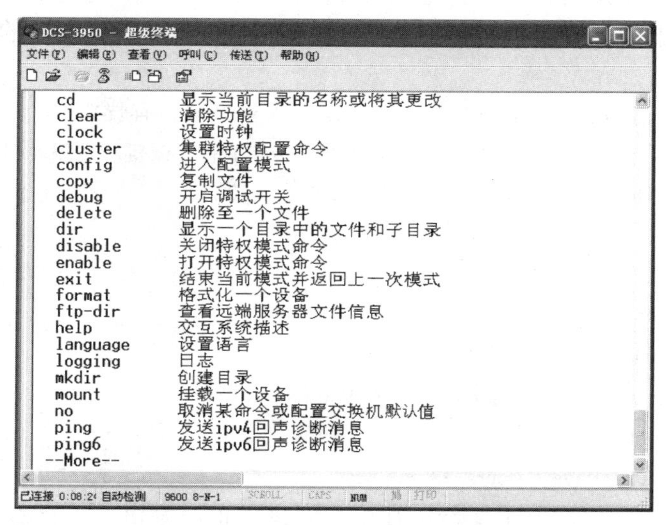

图　1-11

从图 1-11 看到列出了很多特权用户配置模式下的命令。最后一行的“—More—”表示一屏显示不下，后面还有。可以按空格键继续显示下一屏，也可以按<Enter>键继续显示下一行。

除此之外，还有两种场合可以使用“？”。

第一种，如果只记住了某命令的前几个字母，可以使用“？”查询，代码如下。

```
DCS-3950-26C#lan?
  language   设置语言
DCS-3950-26C#co?
  config   进入配置模式
```

```
copy     复制文件
```

表示在特权用户配置模式下以 lan 开头的命令只有一个，而以 co 开头的命令有两个。

第二种，绝大部分命令后需要参数，我们可以用“？”来查询参数，代码如下。

```
DCS-3950-26C(config)#hostname ?
  WORD   主机名 <0-30>字符(符号只允许使用字母、数字和下画线)
  <cr>
```

在命令 hostname 后按空格键后跟一个“？”，可以看到该命令需跟一个主机名，也可以直接按<Enter>键，代码如下。

```
DCS-3950-26C(config)#hostname switch ?
<cr>
DCS-3950-26C(config)#hostname switch
switch(config)#
```

输入 hostname switch 后按空格键和“？”，看到没有什么参数了，可以直接按<Enter>键，然后提示符主机名部分被更改。

任务 5：Tab 键的使用。

学会使用了“？”，就可以不用去记忆繁多的命令了。但 CLI 还提供了更方便快捷的 Tab 键。如果一个命令的前几个字符没有歧义，可以使用 Tab 键来补全命令的输入，代码如下。

```
switch#show ve
% Ambiguous command:   "show ve"          //只输入 ve 不行，有歧义
switch#show ver                           //输入 ver 后按 Tab 键，可以补全命令
  DCS-3950-26C Device, Compiled on Jul 26 16:27:26 2010
  SoftWare Version 6.1.73.13
  BootRom Version 3.0.12
  HardWare Version R01
  Device serial number A830004862
  Copyright (C) 2001-2009 by Digital China Networks Limited.
  All rights reserved
  Uptime is 0 weeks, 0 days, 7 hours, 32 minutes
```

只有当前命令正确的情况下才可以使用 Tab 键。也就是说，一旦命令没有输入完全，但是 Tab 键又没有起作用时，就说明当前的命令中出现了错误，或者命令错误，或者参数错误等，需要仔细排查。

任务 6：命令的不完全匹配。

Tab 键可以帮助我们熟悉命令，而熟练之后，则一般很少使用 Tab 键了。

绝大多数情况下，能使用 Tab 键的地方都可以省略使用它。

例如显示版本号的命令，完全命令是“show version”，使用 Tab 键的代码如下。

```
switch#sh<Tab> ver<Tab>
```

如果把<Tab>键省略，则变成如下代码。

```
switch#sh ver
```

这就是命令的不完全匹配。

任务 7：上下光标键“↑”“↓”的使用。

当输入执行了一些命令后，可以使用上下光标键“↑”“↓”来浏览已执行过的命令，当需要重复执行相似命令时，可以大大节省时间。

任务 8：查看错误信息。

前面已经看到过一个错误信息“%Ambiguous command: "show ve"”，表示命令“show ve”有歧义——可能有两种以上的解释。遇到这种情况，再多输入一个或两个字母，就可避免这种情况。

第二种错误信息是把命令写错了，代码如下。

```
switch>sh valn
^
% Invalid input detected at '^' marker.
```

上例中把“vlan”写成了“valn”，出现了错误信息，并用“^”标示了出错的地方。

第三种错误信息是没有跟命令参数，代码如下。

```
switch#show
% Incomplete command.
```

表示命令不完整。

实训 3　管理交换机配置文件

实训目标

小张：师父，如果交换机配置好了，重启交换机会怎么样呢？配置好的信息保存在哪里？

李工：和我们使用计算机一样，如果没有保存过，重启后配置信息也会丢失。当交换机应用环境发生改变时，就需要清空交换机配置，再重新对交换机进行配置以适应新的应用环境，还需要将已配置的信息进行保存。下面就学习对交换机进行管理的几个任务。

实训拓扑

实训拓扑图如图 1-12 所示。

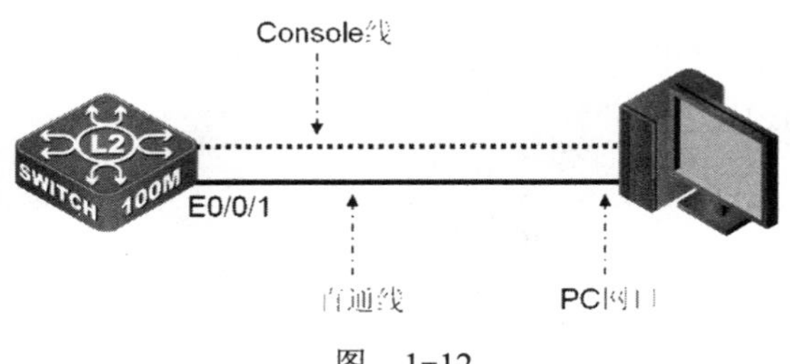

图　1-12

实训任务

任务 1：清空交换机配置。代码如下。

```
switch>enable                               //进入特权用户配置模式
Password:                                   //输入特权密码 digitalchina
switch#set default                          //使用 set default 命令
Are you sure? [Y/N] = y                     //是否确认？
switch#write                                //清空 startup-config 文件
Switch configuration has been set default!
switch #show startup-config                 //显示当前的 startup-config 文件
% Current startup-configuration is default factory configuration!
                                            //系统提示此启动文件为出厂默认配置
switch#reload                               //重新启动交换机
Process with reboot? [Y/N]y
```

小张：这样做了之后，交换机的所有配置信息都清除了吗？特权密码也清除了吗？

李工：是的，你可以试试看。

```
DCS-3950-26C>enable
DCS-3950-26C#
```

小张：果然不需要密码了。

李工：交换机已恢复到出厂配置，所有后来做的配置都清除了。

任务 2：配置交换机日期时间。代码如下。

```
DCS-3950-26C#clock set ?                         //使用？查询命令格式
  HH:MM:SS    Hour:Minute:Second
DCS-3950-26C#clock set 9:52:30                   //配置当前时间
Current time is Sun Jan 01 09:52:30 2006         //配置完即有显示，但年份不对
DCS-3950-26C#clock set 9:52:30 ?                 //使用？查询，原来命令没有结束
  YYYY.MM.DD    Year.Month.Day(Valid time is between 1970.1.1 and 2038.12.31)
<cr>
DCS-3950-26C#clock set 9:52:30 2012.7.31         //配置当前日期和时间
Current time is Tue Jul 31 09:52:30 2012         //显示正确，配置完成
DCS-3950-26C#
```

任务 3：查看交换机闪存中的文件。代码如下。

```
DCS-3950-26C#show flash

nos.img                    4,702,429
startup-config                    24
```

```
Used    4,702,453 bytes in 2 files, Free    3,686,155 bytes.
DCS-3950-26C#
```

可以看到在交换机闪存中保存了两个文件，一个是交换机操作系统文件 nos.img；另一个是配置文件 startup-config。下面来备份这两个文件。

任务 4：配置 TFTP 服务器。

首先，需要把 PC 配置成 TFTP 服务器。TFTP（Trivial File Transfer Protocol，微型文件传输协议）与 FTP 不同，TFTP 采用 UDP 进行文件传输，UDP 在传输时使用的是不可靠的数据流传输服务，同时也不支持用户认证机制以及设置用户对文件操作的授权；它是通过发送包文，应答方式，加上超时重传方式来保证数据的正确传输。TFTP 相对于 FTP 的优点是提供简单的、开销不大的文件传输服务。

市场上 TFTP 服务器的软件很多，各种软件虽然界面不同，但功能都一样，使用方法也都类似：首先是 TFTP 软件安装（有些软件连安装都不需要），安装完毕设定根目录，需要使用的时候，开启 TFTP 服务器即可。下面以 Tftpd32 为例来配置 TFTP 服务器。

双击 Tftpd32.exe 文件，出现 TFTP 主界面，如图 1-13 所示。

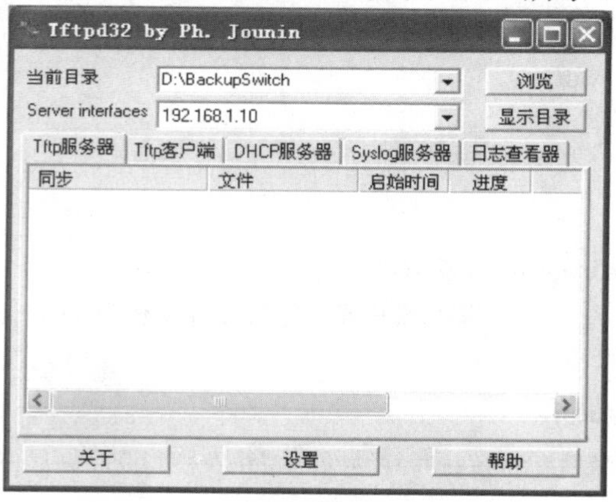

图　1-13

在主界面中可以看到该服务器的根目录是 D:\BackupSwitch，服务器的 IP 地址 192.168.1.10 也自动出现在 Server interfaces 下拉列表中。可以通过单击“设置”按钮来更改 TFTP 服务器的根目录。

任务 5：配置交换机管理 IP 地址。

交换机管理 VLAN 默认为 Vlan 1，可以通过配置 Vlan 1 的 IP 地址实现 PC 网口与交换机以太网口的通信。代码如下。

```
DCS-3950-26C#config terminal                        //进入全局配置模式
DCS-3950-26C(config)#interface vlan 1               //进入 Vlan 1 配置接口
DCS-3950-26C(config-if-vlan1)#ip address 192.168.1.1 255.255.255.0
                                                    //配置 Vlan 1 的 IP 地址与掩码
DCS-3950-26C(config-if-vlan1)#no shutdown           //打开 Vlan 1 接口
DCS-3950-26C(config-if-vlan1)#exit                  //返回全局配置模式
DCS-3950-26C(config)#
```

任务6：验证TFTP服务器与交换机的连通性。代码如下。

```
DCS-3950-26C#ping 192.168.1.10                    //特权模式下 ping TFTP 服务器 IP
Type ^c to abort.
Sending 5 56-byte ICMP Echos to 192.168.1.10, timeout is 2 seconds.
!!!!!                                              //出现 5 个“!”号，表示 ping 通了
Success rate is 100 percent (5/5), round-trip min/avg/max = 0/3/16 ms
DCS-3950-26C#
```

任务7：备份交换机配置文件。

将任务3中看到的交换机配置文件备份到TFTP服务器中。代码如下。

```
DCS-3950-26C#write                                 //保存当前配置到 startup-config 中
Write configuration successfully!
DCS-3950-26C#copy startup-config tftp://192.168.1.10/sc2012.7
                    //将配置文件 startup-config 上传到 TFTP 服务器根目录中，并改名
                      为 sc2012.7
Confirm copy file [Y/N]:y
Begin to send file, please wait...

File transfer complete.
close tftp client.
DCS-3950-26C#copy nos.img tftp://192.168.1.10
                    //将交换机操作系统文件上传到 TFTP 服务器根目录中，不改名
Confirm copy file [Y/N]:y
Begin to send file, please wait...
################################################################################
################################################
File transfer complete.
close tftp client.
DCS-3950-26C#
```

任务8：下载配置文件。代码如下。

```
DCS-3950-26C#copy tftp://192.168.1.10/sc2012.7 startup-config
              //将备份文件 sc2012.7 恢复成交换机启动配置文件 startup-config
Confirm to overwrite the existed destination file?  [Y/N]:y
Begin to receive file, please wait...

File transfer complete.
Recv total 907 bytes
Write ok.
close tftp client.
```

```
DCS-3950-26C#reload                    //重启交换机
```

任务 9：升级交换机操作系统。

下载交换机操作系统升级版本到 TFTP 服务器根目录，代码如下。

```
DCS-3950-26C#copy tftp://192.168.1.10/nos.img nos.img
Confirm to overwrite the existed destination file?   [Y/N]:y
Begin to receive file, please wait...
Get Img file size success, Img file size is:4702429(bytes).
##########################################################################################
##################################
File transfer complete.
Recv total 4702429 bytes

Write ok.
close tftp client.
DCS-3950-26C#
```

实训 4 交换机带内管理

实训目标

小张：师父，如果一个局域网内有很多交换机分别放置在不同的位置，这样一台一台地去配置不是很麻烦吗？

李工：我们前面学习的是交换机的带外管理方式，也就是使用交换机的 Console 口去管理。而网络是互通的，在网络的任何一个信息点都应该能访问其他的信息点，为什么不通过网络的方式来管理交换机呢？通过带内管理方式，管理员就可以在办公室中调试网络中所有的交换机了。

带内管理方式有两种，分别是 Telnet 方式和 Web 方式。

但是，如果交换机的某些配置或错误导致带内管理方式失效，就必须使用带外管理对交换机进行配置管理。

实训拓扑

实训拓扑图如图 1-14 所示。

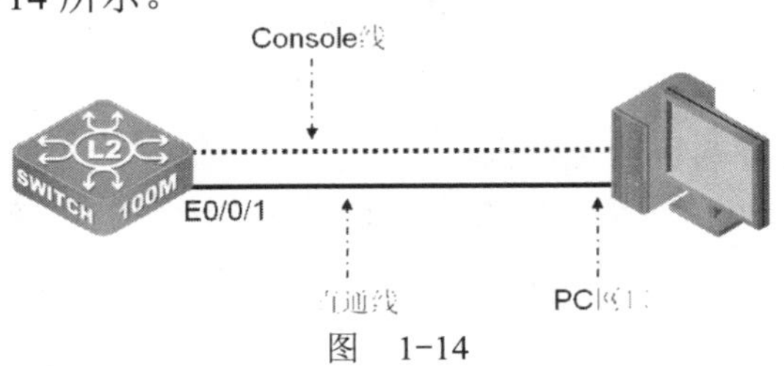

图 1-14

实训任务

任务 1：通过 Console 口配置交换机的基本信息。

在使用带内管理方式前，必须对交换机做一些基本配置。代码如下。

```
//配置特权用户模式密码为 digitalchina
DCS-3950-26C>enable
DCS-3950-26C#config terminal
DCS-3950-26C(config)#enable password digitalchina
DCS-3950-26C(config)#

//配置用户名认证，用户名为 dcs，密码为 digital，权限为 15（管理员）
DCS-3950-26C(config)#username dcs privilege 15 password digital
DCS-3950-26C(config)#

//开启 Telnet 服务和 Web 服务
DCS-3950-26C(config)#telnet-server enable
Telnet server has been already enabled.                //Telnet 服务默认是开启的
DCS-3950-26C(config)#ip http server
web server is on
DCS-3950-26C(config)#

//配置管理 Vlan 的 IP 地址
DCS-3950-26C(config)#interface vlan 1
DCS-3950-26C(config-if-vlan1)#ip address 192.168.1.1 255.255.255.0
DCS-3950-26C(config-if-vlan1)#exit
DCS-3950-26C(config)#
```

任务 2：验证 PC 与交换机的连通性。代码如下。

```
DCS-3950-26C#ping 192.168.1.10
Type ^c to abort.
Sending 5 56-byte ICMP Echos to 192.168.1.10, timeout is 2 seconds.
!!!!!                              //出现 5 个“!”，表示交换机与 PC 通信正常
Success rate is 100 percent (5/5), round-trip min/avg/max = 0/0/0 ms
DCS-3950-26C#
```

任务 3：使用 Telnet 方式管理交换机。

此时，可以拔除 Console 连接。单击“开始”→“运行”命令，在“运行”对话框中输入“cmd”，启动命令提示符窗口，输入 telnet 192.168.1.1，使用 Telnet 方式登录交换机。如图 1-15 所示。

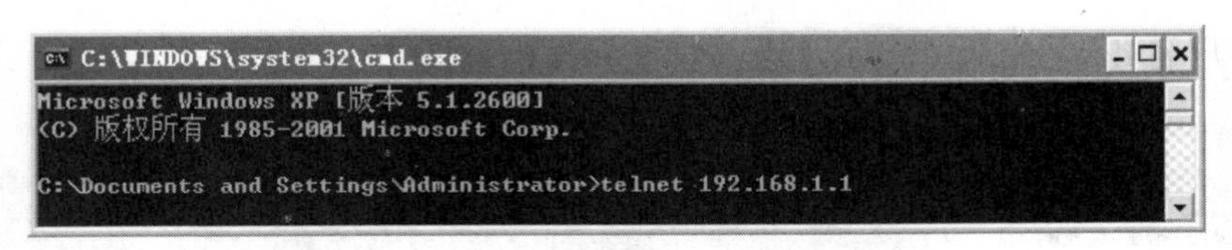

图　1-15

输入用户名 dcs，密码 digital，便进入了交换机特权用户配置模式（用户 dcs 权限等级为最高级 15，直接进入特权用户配置模式），如图 1-16 所示。

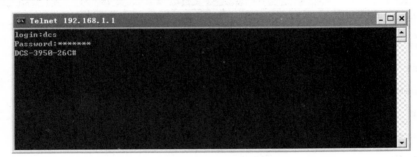

图　1-16

任务 4：使用 Web 方式管理交换机。

打开浏览器，在地址栏输入 http://192.168.1.1，便出现了交换机的 Web 登录界面，如图 1-17 所示。

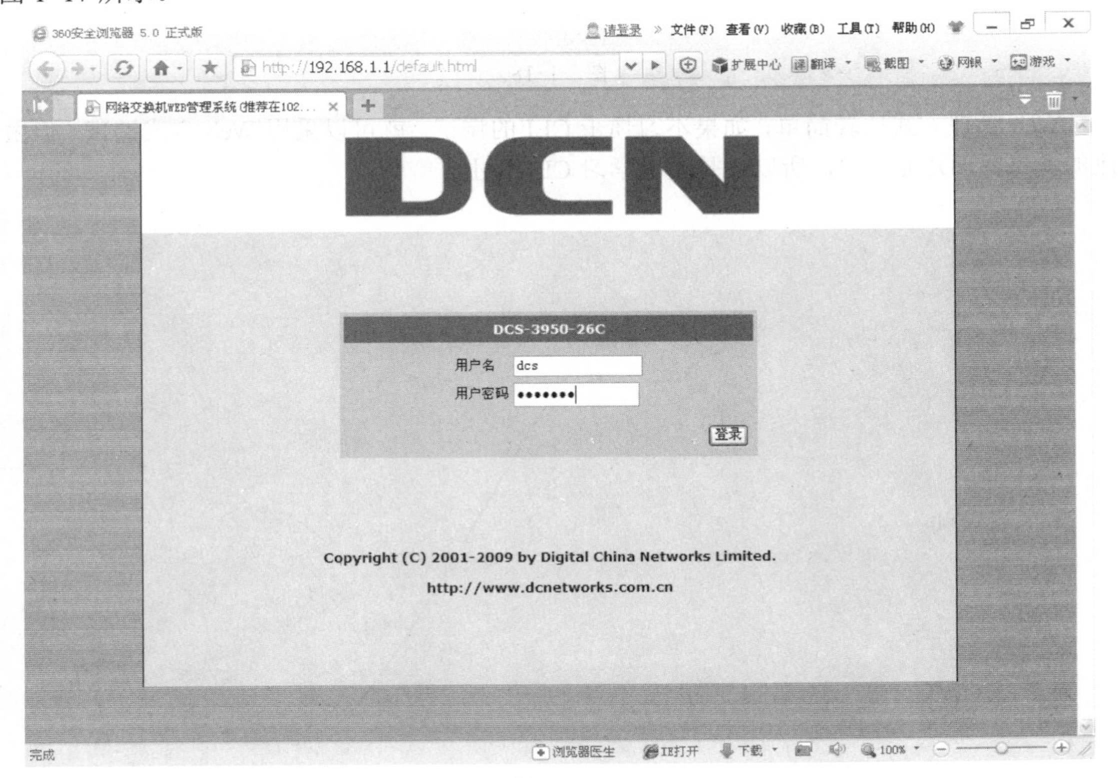

图　1-17

输入用户名 dcs，密码 digital，进入管理界面，如图 1-18 所示。

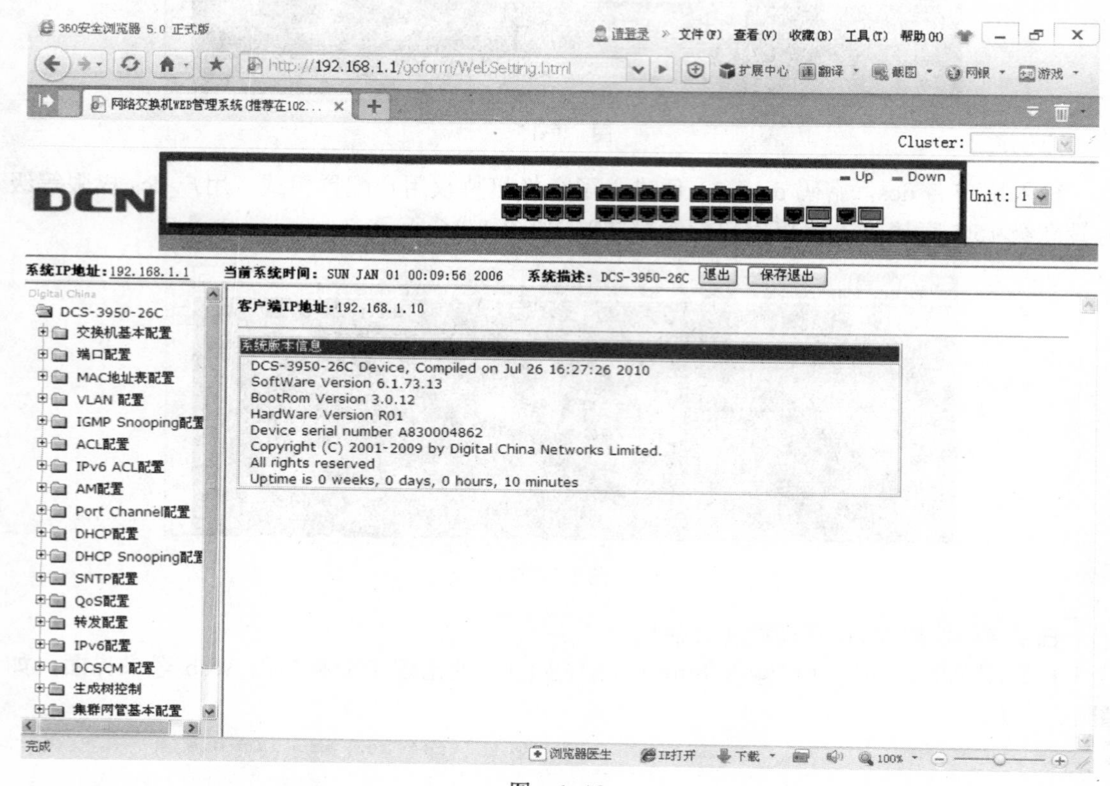

图 1-18

Web 管理方式比较简单，如果不习惯于 CLI 的调试，就可以采用 Web 方式调试。但主流的调试界面还是 CLI，所以还是要以学习 CLI 为主。

第2章　交换机典型园区应用实践

实训1　单台交换机 VLAN

实训目标

李工：现在我们开始进入园区网的应用实践。首先就要接触 VLAN。小张，还记得什么是 VLAN 以及为什么要划分 VLAN 吗？

小张：VLAN 是虚拟局域网的意思，划分 VLAN 的主要目的是隔离广播域。

李工：对。由于交换机所有端口属于同一个广播域，而网络中存在大量的广播包，VLAN 技术能够有效地进行隔离，从而使网络性能得到提升。

例如某学校的两个机房位于同一楼层，所有信息端口都连接在一台交换机上。为了保证两个机房的相对独立，不影响各机房内部的通信效率，保证两个机房之间互不干扰，就需要划分 VLAN，使这两个机房处于不同的 VLAN 内。

实训拓扑

实训拓扑图如图 2-1 所示。

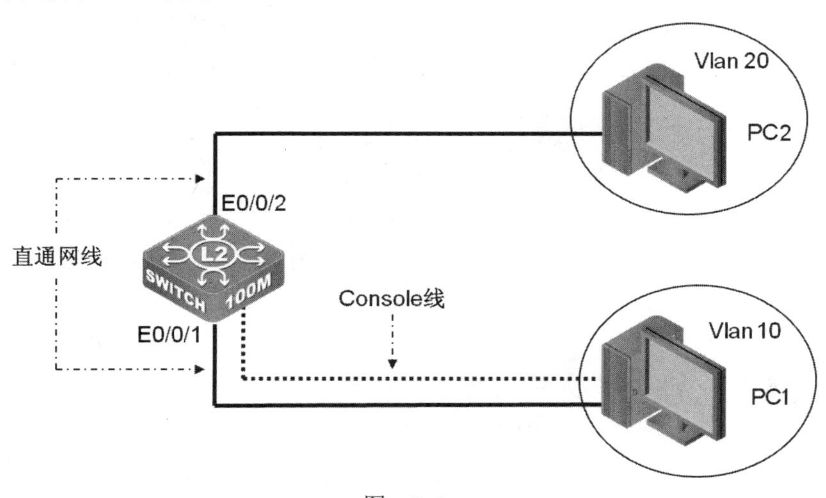

图　2-1

实训任务

任务 1：规划交换机的 VLAN 及端口，分配 PC 的 IP 地址，见表 2-1。

表 2-1

VLAN	交换机端口	设备	IP 地址	掩码
10	0/0/1-0/0/5	PC1	192.168.1.101	255.255.255.0
20	0/0/6-0/0/10	PC2	192.168.1.102	255.255.255.0

任务 2：在交换机上创建 VLAN 并验证。代码如下。

```
DCS-3950-26C>enable
Password:                                  //输入特权密码 digitalchina
DCS-3950-26C#config terminal               //进入全局配置模式
DCS-3950-26C(config)#vlan 10               //创建 vlan10
DCS-3950-26C(config-vlan10)#exit           //返回全局模式
DCS-3950-26C(config)#vlan 20               //创建 vlan20
DCS-3950-26C(config-vlan20)#exit
```

验证配置，代码如下。

```
DCS-3950-26C#show vlan
VLAN Name         Type       Media     Ports
---- ------------ ---------- --------- ----------------------------------------
1    default      Static     ENET      Ethernet0/0/1      Ethernet0/0/2
                                       Ethernet0/0/3      Ethernet0/0/4
                                       Ethernet0/0/5      Ethernet0/0/6
                                       Ethernet0/0/7      Ethernet0/0/8
                                       Ethernet0/0/9      Ethernet0/0/10
                                       Ethernet0/0/11     Ethernet0/0/12
                                       Ethernet0/0/13     Ethernet0/0/14
                                       Ethernet0/0/15     Ethernet0/0/16
                                       Ethernet0/0/17     Ethernet0/0/18
                                       Ethernet0/0/19     Ethernet0/0/20
                                       Ethernet0/0/21     Ethernet0/0/22
                                       Ethernet0/0/23     Ethernet0/0/24
                                       Ethernet0/0/25     Ethernet0/0/26
10   VLAN0010     Static     ENET
20   VLAN0020     Static     ENET
DCS-3950-26C#
```

可以看到已经创建了 VLAN，但 VLAN 中还没有端口。

任务 3：给 VLAN 添加端口。代码如下。

```
DCS-3950-26C(config)#vlan 10
DCS-3950-26C(config-vlan10)#switchport interface ethernet 0/0/1-5
Set the port Ethernet0/0/1 access vlan 10 successfully
Set the port Ethernet0/0/2 access vlan 10 successfully
Set the port Ethernet0/0/3 access vlan 10 successfully
Set the port Ethernet0/0/4 access vlan 10 successfully
Set the port Ethernet0/0/5 access vlan 10 successfully
DCS-3950-26C(config-vlan10)#exit
DCS-3950-26C(config)#vlan 20
DCS-3950-26C(config-vlan20)#switchport interface ethernet 0/0/6-10
Set the port Ethernet0/0/6 access vlan 20 successfully
Set the port Ethernet0/0/7 access vlan 20 successfully
Set the port Ethernet0/0/8 access vlan 20 successfully
Set the port Ethernet0/0/9 access vlan 20 successfully
Set the port Ethernet0/0/10 access vlan 20 successfully
DCS-3950-26C(config-vlan20)#exit
DCS-3950-26C(config)#
```

验证配置，代码如下。

```
DCS-3950-26C#show vlan
VLAN Name         Type       Media      Ports
---- ----------- ---------- --------- ----------------------------------------
1    default      Static     ENET       Ethernet0/0/11       Ethernet0/0/12
                                        Ethernet0/0/13       Ethernet0/0/14
                                        Ethernet0/0/15       Ethernet0/0/16
                                        Ethernet0/0/17       Ethernet0/0/18
                                        Ethernet0/0/19       Ethernet0/0/20
                                        Ethernet0/0/21       Ethernet0/0/22
                                        Ethernet0/0/23       Ethernet0/0/24
                                        Ethernet0/0/25       Ethernet0/0/26
10   VLAN0010     Static     ENET       Ethernet0/0/1        Ethernet0/0/2
                                        Ethernet0/0/3        Ethernet0/0/4
                                        Ethernet0/0/5
20   VLAN0020     Static     ENET       Ethernet0/0/6        Ethernet0/0/7
                                        Ethernet0/0/8        Ethernet0/0/9
                                        Ethernet0/0/10
DCS-3950-26C#
```

任务 4：验证互通性。见表 2-2。

表 2-2

PC1 位置	PC2 位置	动　作	结　果	备　注
0/0/1	0/0/2	PC1 ping PC2	通	PC1 与 PC2 同在 VLAN10 中
0/0/6	0/0/7	PC1 ping PC2	通	PC1 与 PC2 同在 VLAN20 中
0/0/1	0/0/6	PC1 ping PC2	不通	PC1 与 PC2 不在同一 VLAN 中
0/0/8	0/0/3	PC1 ping PC2	不通	PC1 与 PC2 不在同一 VLAN 中

实训 2　跨交换机 VLAN

实训目标

小张：师父，我们部门在二楼和三楼都有办公室，也是用一台交换机连接吗？

李工：当然不行。这样做布线很困难，并且一台盒式交换机一般只有 24 或 48 个接口，如果用户数量超过端口限制，就需要增加交换机来实现用户接入。这种情况下，我们可以使用跨交换机的 VLAN 来实现多台交换机间用户数据的互联互通。

实训拓扑

实训拓扑图如图 2-2 所示。

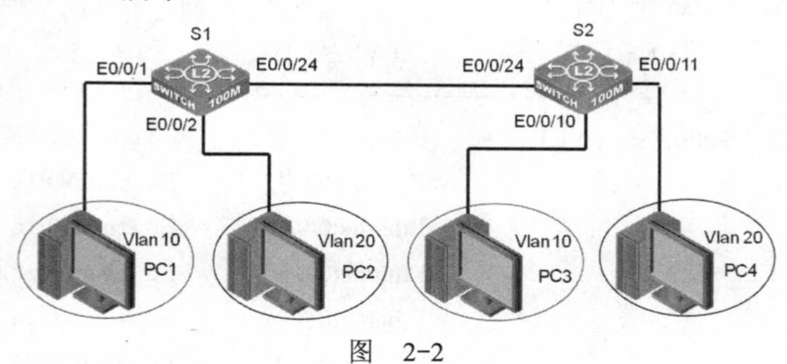

图 2-2

实训任务

任务 1：规划交换机的 VLAN 及端口，分配 PC 的 IP 地址。见表 2-3。

表 2-3

VLAN	交换机端口	设　备	IP 地 址	掩　码
10	S1 0/0/1	PC1	192.168.1.10	255.255.255.0
	S2 0/0/10	PC3	192.168.1.11	255.255.255.0
20	S1 0/0/2	PC2	192.168.2.10	255.255.255.0
	S2 0/0/11	PC4	192.168.2.11	255.255.255.0
Trunk 口	S1 0/0/24			
	S2 0/0/24			

任务 2：理解交换机的端口类型。

小张：师父，什么是 Trunk 口？

李工：交换机常见的端口类型有 Access 端口和 Trunk 端口。Access 端口主要用来接入终端设备，如 PC、服务器等。这种端口只能承载一个 Vlan 的流量。当它接收到一个来自终端设备的以太网帧时，会打上自己的 PVID（Port Vlan ID）然后转发，如图 2-3 所示。当 Access 端口向终端设备发送数据时，先判断数据帧的 VID 是否和自己的 PVID 一致，如果一致，则剥离 VID 信息转发给终端设备，否则丢弃，如图 2-4 所示。

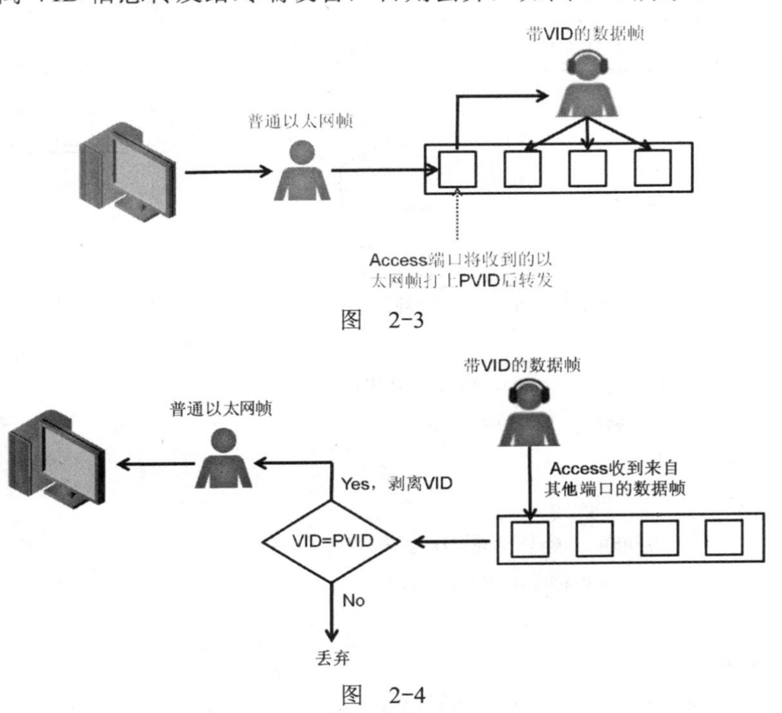

图　2-3

图　2-4

Trunk 端口主要用于交换机之间或交换机与上层网络设备之间的连接，它可以承载多个 Vlan 的流量。当它接收到一个来自 Access 端口的数据帧时，如果此数据帧的 VID 与自己的 PVID 一致，则剥离 VID 信息后转发到干道链路。如果不一致，则直接转发至干道链路。当它接收到来自干道链路的数据帧时，如果该数据帧无 VID 信息，则打上自己的 PVID 后转发至 Access 端口。如果有 VID 信息，则直接转发至 Access 端口。如图 2-5 所示。

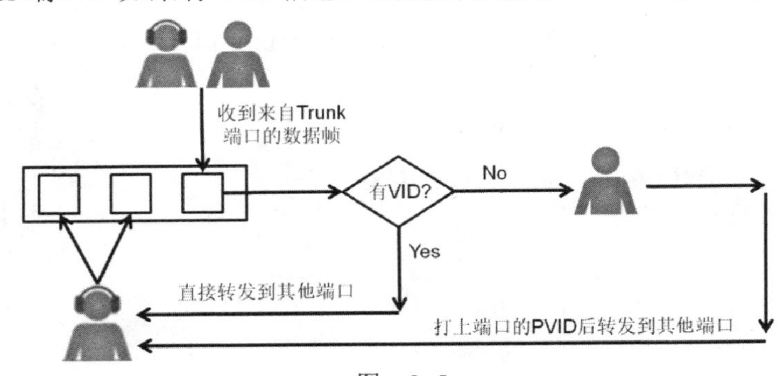

图　2-5

除了 Trunk 和 Access 以外，还有第三种 Hybrid 的端口类型。Hybrid 端口类似于 Trunk 端口，可以透传多个 Vlan。但和 Trunk 端口的区别是，Hybrid 允许透传的 Vlan 中有多个 Vlan 不打 PVID。当 Hybrid 允许多个 Vlan 不打 PVID 时，收到一个没有 PVID 的报文依然使用端口默认的 PVID（Hybrid 端口的默认 PVID 需要手动指定）进行转发。如图 2-6 所示。

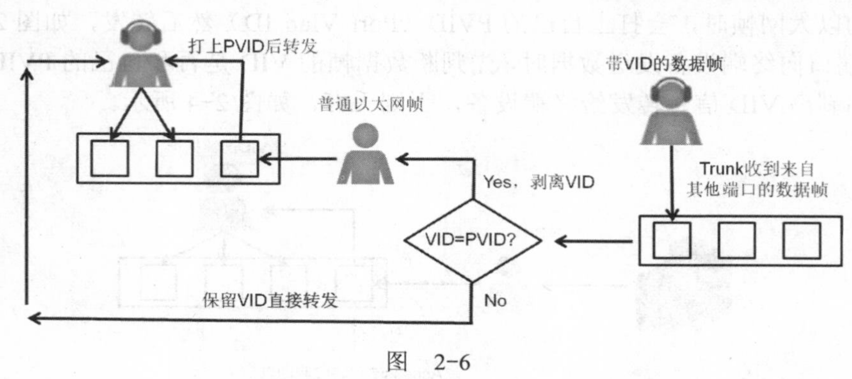

图 2-6

任务 3：在交换机 S1 中划分 Vlan、分配端口及 Trunk 口。代码如下。

```
S1(config)#vlan 10
S1(config-vlan10)#switchport interface ethernet 0/0/1
Set the port Ethernet0/0/1 access vlan 10 successfully
S1(config-vlan10)#exit
S1(config)#vlan 20
S1(config-vlan20)#switchport interface ethernet 0/0/2
Set the port Ethernet0/0/2 access vlan 20 successfully
S1(config-vlan20)#exit
S1(config)#interface ethernet 0/0/24
S1(config-if-ethernet0/0/24)#switchport mode trunk
Set the port Ethernet0/0/24 mode Trunk successfully
S1(config-if-ethernet0/0/24)#switchport trunk allowed vlan all
set the Trunk port Ethernet0/0/24 allowed vlan successfully
S1(config-if-ethernet0/0/24)#exit
S1(config)#
```

任务 4：在交换机 S2 中划分 Vlan、分配端口及 Trunk 口。代码如下。

```
S2(config)#vlan 10
S2(config-vlan10)#switchport interface ethernet 0/0/10
Set the port Ethernet0/0/10 access vlan 10 successfully
S2(config-vlan10)#exit
S2(config)#vlan 20
S2(config-vlan20)#switchport interface ethernet 0/0/11
Set the port Ethernet0/0/11 access vlan 20 successfully
S2(config-vlan20)#exit
```

```
S2(config)#interface ethernet 0/0/24
S2(config-if-ethernet0/0/24)#switchport mode trunk
Set the port Ethernet0/0/24 mode Trunk successfully
S2(config-if-ethernet0/0/24)#switchport trunk allowed vlan all
set the Trunk port Ethernet0/0/24 allowed vlan successfully
S2(config-if-ethernet0/0/24)#exit
S2(config)#
```

任务 5：验证互通性。见表 2-4。

表　2-4

源主机	端口	目标主机	端口	动作	结果
PC1	S1 0/0/1	PC2	S1 0/0/2	Ping	不通
		PC3	S2 0/0/10	Ping	通
		PC4	S2 0/0/11	Ping	不通

实训 3　生成树协议

实训目标

李工：在交换网络中，为避免因环路而产生广播风暴和交换设备地址系统失效问题，同时使用冗余链路提高网络的可靠性，交换机提供了生成树协议来解决这些问题。

小张：广播风暴？地址失效？冗余链路？

李工：所谓广播风暴，是指在交换网络中每台交换机不停地从每个端口将同一个数据转发出去，当从某一端口收到时又再一次进行所有端口的转发，这样，交换机的带宽将被耗尽而无法对正常数据进行转发，从而形成广播风暴。

在交换机中，每一台所连接设备的 MAC 地址与所连接的端口号形成的对应表称为 MAC 地址表。伴随着广播风暴的出现，交换机中的 MAC 地址表会出现混乱与错误，加深了广播风暴的影响。

交换网络中使物理上形成环路的同时逻辑上阻断环路，形成冗余链路来提高网络的可靠性，这就是生成树协议。

实训拓扑

实训拓扑图如图 2-7 所示。

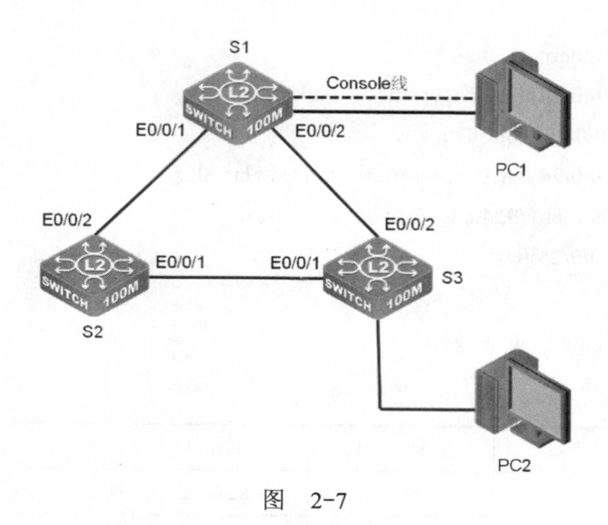

图 2-7

实训任务

任务 1：了解广播风暴的危害。

设置 PC1 的 IP 地址为 192.168.1.10，PC2 的 IP 地址为 192.168.1.11，PC1 ping PC2，结果如下。

```
C:\Documents and Settings\Administrator>ping 192.168.1.11

Pinging 192.168.1.11 with 32 bytes of data:

Request timed out.
Request timed out.
Request timed out.
Request timed out.

Ping statistics for 192.168.1.11:
    Packets: Sent = 4, Received = 4, Lost = 0 (0% loss),
Approximate round trip times in milli-seconds:
    Minimum = 1ms, Maximum = 4ms, Average = 1ms
```

由于生成树协议在神码交换机中默认是关闭的，可以看到此时已无法 Ping 通，交换机 Link 指示灯频繁闪烁，PC 反应迟缓，整个网络已被广播风暴袭击，无法进行正常通信。

任务 2：理解生成树协议。

生成树协议要解决的问题是，在网络中有物理环路的情况下，从逻辑上阻断环路，避免广播风暴的发生，同时，在链路出现故障时，能够自动启用冗余链路恢复通信。

IEEE 802.1d 协议通过在交换机上运行一套复杂的算法 STA（Spanning-tree algorithm），使冗余端口置于“阻断状态”，使得接入网络的计算机在与其他计算机通信时，只有一条链路生效，而当这个链路出现故障无法使用时，IEEE 802.1d 协议会重新计算网络链路，将处于“阻断状态”的端口重新打开，从而既保障了网络正常运转，又保证了冗余能力。

生成树协议通过选择根桥、根端口、指定端口 3 个步骤来完成生成树的计算。

任务 3：选择根网桥。

根网桥的选择是通过交换机的网桥 ID 来确定的。打开交换机的生成树协议，并查询交换机的网桥 ID。代码如下。

```
S1(config)#spanning-tree                                   //开启生成树协议

MSTP is starting now, please wait.............
MSTP is enabled successfully.
S1(config)#exit
S1#show spanning-tree

                 -- MSTP Bridge Config Info --

Standard       :   IEEE 802.1s
Bridge MAC     :   00:03:0f:0f:2a:63
Bridge Times :   Max Age 20, Hello Time 2, Forward Delay 15
Force Version:   3

########################### Instance 0 ###########################
Self Bridge Id      : 32768 -    00:03:0f:0f:2a:63
Root Id             : this switch
Ext.RootPathCost : 0
Region Root Id      : this switch
Int.RootPathCost : 0
Root Port ID        : 0
Current port list in Instance 0:
Ethernet0/0/1 Ethernet0/0/2 (Total 2)

PortName       ID       ExtRPC     IntRPC   State Role    DsgBridge           DsgPort

-------------- -------- --------- --------- --- ---- ------------------ -------

Ethernet0/0/1 128.001         0           0 FWD   DSGN 32768.00030f0f2a63 128.001
Ethernet0/0/2 128.002         0           0 FWD   DSGN 32768.00030f0f2a63 128.002
S1#
```

其中，Self Bridge Id: 32768 -　00:03:0f:0f:2a:63 就是该交换机的网桥 ID。它由两个部分组成，前面的数字 32768 是交换机的默认优先级，其取值范围为 0～65535，步长为 4096，数字越小则优先级越高；后面是该交换机的 MAC 地址。若各交换机优先级相同，则比较交换机的 MAC 地址，越小则优先级越高。

同样可以查询到 S2 的网桥 ID 为 Self Bridge Id:32768 - 00:03:0f:0f:2a:7d，交换机 S3 的网桥 ID 为 Self Bridge Id:32768 - 00:03:0f:0f:1c:8e。

比较三个网桥 ID 的大小，它们的优先级相同，而 MAC 地址中最小的是交换机 S3。因此，S3 被生成树协议选定为根网桥。

任务 4：选择根端口。

根网桥确定后，需要在非根网桥上选择根端口。选择根端口的依据是：

1）端口到达根网桥的路径成本最低。路径成本是指从非根网桥到根网桥上所有链路的成本之和。神码设备默认 10Mbit/s/100Mbit/s 自适应的路径开销为 200000。

2）若路径成本无法选择根端口，则依据非根网桥的直连（上游）网桥的桥 ID 最小来选择根端口，这在交换机级联时可能出现此种情况。

3）若依然无法确定根端口，则再依据上游端口 ID 最小来确定根端口。端口 ID 形如 128.001，前面的 128 为端口优先级，其取值范围为 0～255，步长为 1，默认值为 128。后面的 001 为端口编号。

根据以上依据，在此例中，S3 为根网桥，我们需要在非根网桥 S1 和 S2 中选择根端口。

S1 端口 E0/0/1 路径成本=200000+200000=400000；

S1 端口 E0/0/2 路径成本=200000；

S2 端口 E0/0/1 路径成本=200000；

S2 端口 E/0/0/2 路径成本=200000+200000=400000；

所以，S1 的 E/0/0/2 和 S2 的 E/0/0/1 端口为根端口。

任务 5：选择指定端口。

每条连接交换机的链路上都需要选择指定端口。根网桥的所有端口都是指定端口，非根网桥的指定端口依据以下方法选择：

1）根路径成本最低。

2）端口所在网桥的桥 ID 值较小。

3）端口 ID 值较小。

在此例中，S3 为根网桥，其所有端口都是指定端口，S1 的 E0/0/2 和 S2 的 E0/0/1 为根端口，所以，我们需要在 S1 的 E0/0/1 和 S2 的 E0/0/2 中选择指定端口。

先比较根路径成本：

S1 的 E0/0/1 到根网桥的路径成本：经过 S2 交换机连接到 S3，故路径成本为 400000，同样，S2 的 E/0/0/2 根路径成本也是 400000，无法比较。

再比较端口所在网桥的桥 ID 值。如前所述，S1 的桥 ID 为 32768 - 00:03:0f:0f:2a:63，S2 的桥 ID 为 32768 - 00:03:0f:0f:2a:7d，S1 的桥 ID 值较小，所以 S1 的 E0/0/1 端口为指定端口。

最后，将既非根端口又非指定端口的 S2 E0/0/2 端口关闭，生成树协议计算完成，封闭了物理环路，并将 S1 的 E0/0/1 到 S2 的 E0/0/2 链路作为冗余链路。

任务 6：验证生成树的冗余性。

李工：小张，你来验证下。

小张：好的。我用 PC1 ping PC2

代码如下。

```
C:\Documents and Settings\Administrator>ping 192.168.1.11

Pinging 192.168.1.11 with 32 bytes of data:

Reply from 192.168.1.11: bytes=32 time=4ms TTL=64
Reply from 192.168.1.11: bytes=32 time=1ms TTL=64
Reply from 192.168.1.11: bytes=32 time=1ms TTL=64
Reply from 192.168.1.11: bytes=32 time=1ms TTL=64

Ping statistics for 192.168.1.11:
    Packets: Sent = 4, Received = 4, Lost = 0 (0% loss),
Approximate round trip times in milli-seconds:
    Minimum = 1ms, Maximum = 4ms, Average = 1ms
```

可以正常通信，没有广播风暴发生，生成树协议工作正常。

查看 S2 的生成树协议信息。代码如下。

```
S2#show spanning-tree

                    -- MSTP Bridge Config Info --

Standard        :  IEEE 802.1s
Bridge MAC      :  00:03:0f:0f:2a:7d
Bridge Times :  Max Age 20, Hello Time 2, Forward Delay 15
Force Version:  3

########################### Instance 0 ###########################
Self Bridge Id     : 32768 -  00:03:0f:0f:2a:7d
Root Id            : 32768 -  00:03:0f:0f:1c:8e
Ext.RootPathCost : 200000
Region Root Id     : this switch
Int.RootPathCost : 0
Root Port ID       : 128.1
Current port list in Instance 0:
Ethernet0/0/1 Ethernet0/0/2 (Total 2)

PortName      ID      ExtRPC    IntRPC   State Role      DsgBridge          DsgPort

------------- -------- --------- --------- ---  ---- ------------------ -------
```

```
Ethernet0/0/1 128.001        0          0 FWD   DSGN 32768.00030f0f2a7d 128.001
Ethernet0/0/2 128.002        0          0 BLK BKUP 32768.00030f0f2a7d 128.002
```

看到 S2 的 E0/0/2 端口处于阻塞状态。

把 S1 到 S3 的链路拔除，再用 PC1 ping PC2。代码如下。

```
C:\Documents and Settings\Administrator>ping 192.168.1.11

Pinging 192.168.1.11 with 32 bytes of data:

Reply from 192.168.1.11: bytes=32 time=2ms TTL=64
Reply from 192.168.1.11: bytes=32 time=1ms TTL=64
Reply from 192.168.1.11: bytes=32 time=12ms TTL=64
Reply from 192.168.1.11: bytes=32 time=1ms TTL=64

Ping statistics for 192.168.1.11:
    Packets: Sent = 4, Received = 4, Lost = 0 (0% loss),
Approximate round trip times in milli-seconds:
    Minimum = 1ms, Maximum = 12ms, Average = 4ms
```

仍然可以正常通信，再次查看 S2 的生成树协议。代码如下。

```
S2#show spanning-tree

                -- MSTP Bridge Config Info --

Standard       :  IEEE 802.1s
Bridge MAC     :  00:03:0f:0f:2a:7d
Bridge Times :  Max Age 20, Hello Time 2, Forward Delay 15
Force Version:  3

########################### Instance 0 ###########################
Self Bridge Id     : 32768 -   00:03:0f:0f:2a:7d
Root Id             : 32768 -   00:03:0f:0f:1c:8e
Ext.RootPathCost : 200000
Region Root Id     : this switch
Int.RootPathCost : 0
Root Port ID       : 128.1
Current port list in Instance 0:
Ethernet0/0/1 Ethernet0/0/2 (Total 2)

PortName     ID     ExtRPC     IntRPC   State Role     DsgBridge        DsgPort
```

```
-------------- -------- --------- --------- ---    ---- ------------------ -------

Ethernet0/0/1 128.001         0              0 FWD    DSGN 32768.00030f0f2a7d 128.001
Ethernet0/0/2 128.002         0              0 FWD    DSGN 32768.00030f0f2a7d 128.002
```

看到 E0/0/2 端口已处于转发状态，冗余链路工作正常。

任务 7：修改优先级改变生成树形态。

李工：前面使用了交换机及端口的默认优先级来运行生成树协议。在有些情况下，我们需要特别指定某台交换机为根网桥，或特别指定某端口为指定端口时，可以采用改变交换机或端口的优先级来达到目的。

在本例中，如果指定 S1 为根网桥，则需要以 4096 为倍数减小 S1 的优先级值，使 S1 获得最高优先权。

```
S1(Config)#spanning-tree mst 0 priority 28672                  //改变 S1 交换机优先级
```

查看 S2 交换机生成树协议。代码如下。

```
S2#show spanning-tree

                      -- MSTP Bridge Config Info --

Standard          :   IEEE 802.1s
Bridge MAC        :   00:03:0f:0f:2a:7d
Bridge Times :   Max Age 20, Hello Time 2, Forward Delay 15
Force Version:    3

############################ Instance 0 ############################
Self Bridge Id      : 32768 -    00:03:0f:0f:2a:7d
Root Id             : 28672 -    00:03:0f:0f:2a:63
Ext.RootPathCost : 200000
Region Root Id      : this switch
Int.RootPathCost : 0
Root Port ID        : 128.2
Current port list in Instance 0:
Ethernet0/0/1 Ethernet0/0/2 (Total 2)

PortName        ID       ExtRPC     IntRPC   State Role        DsgBridge          DsgPort

-------------- -------- --------- --------- ---    ---- ------------------ -------

Ethernet0/0/1 128.001         0              0 BLK    BKUP 32768.00030f0f2a7d 128.001
Ethernet0/0/2 128.002         0              0 FWD    DSGN 32768.00030f0f2a7d 128.002
```

我们看到 S2 的端口 E0/0/1 被阻塞，E0/0/2 成为根端口，生成树形态已被改变。

实训 4　多实例生成树协议

实训目标

李工：相对于基本生成树，多实例生成树允许多个具有相同拓扑的 VLAN 映射到一个生成树实例上，而这个生成树拓扑同其他生成树实例相互独立。这种机制多重生成树实例为映射到它的 VLAN 的数据流量提供了独立的发送路径，实现不同实例间 VLAN 数据流量的负载分担。

多实例生成树由于多个 VLAN 可以映射到一个单一的生成树实例，IEEE 802.1s 委员会提出了 MST 域的概念，用来解决如何判断某个 VLAN 映射到哪个生成树实例的问题。在这个实训环境中可以进一步理解多 VLAN 的生成树协议原理和实际拓扑生成。

实训拓扑

实训拓扑图如图 2-8 所示。

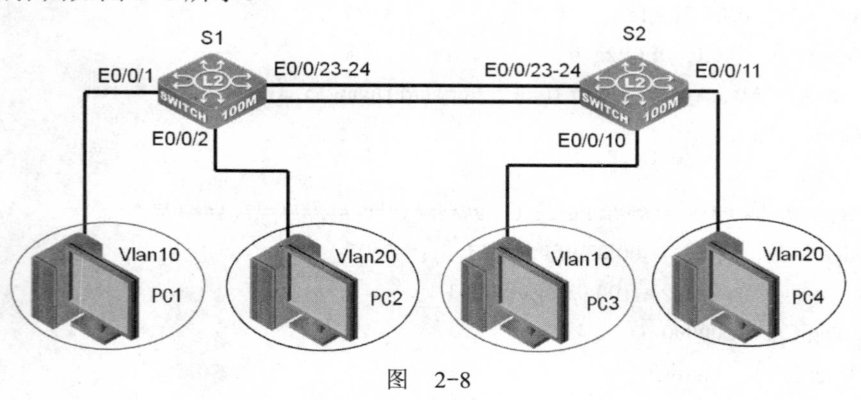

图　2-8

实训任务

任务 1：规划交换机 VLAN 及端口分配，PC 的 IP 地址。见表 2-5。

表　2-5

VLAN	交换机端口	设　　备	IP　地　址	掩　　码
10	S1　0/0/1	PC1	192.168.1.10	255.255.255.0
	S2　0/0/10	PC3	192.168.1.11	255.255.255.0
20	S1　0/0/2	PC2	192.168.2.10	255.255.255.0
	S2　0/0/11	PC4	192.168.2.11	255.255.255.0
Trunk 口	S1　0/0/23-24			
	S2　0/0/23-24			

任务 2：在交换机上划分 VLAN，分配端口及 Trunk 端口。

交换机 S1：

```
S1(config)#vlan 10
S1(config-vlan10)#switchport interface ethernet 0/0/1
Set the port Ethernet0/0/1 access vlan 10 successfully
S1(config-vlan10)#exit
S1(config)#vlan 20
S1(config-vlan20)#switchport interface ethernet 0/0/2
Set the port Ethernet0/0/2 access vlan 20 successfully
S1(config-vlan20)#exit
S1(config)#interface ethernet 0/0/23-24
S1(config-if-port-range)#switchport mode trunk
Set the port Ethernet0/0/23 mode Trunk successfully
Set the port Ethernet0/0/24 mode Trunk successfully
S1(config-if-port-range)#exit
S1(config)#
```

交换机 S2：

```
S1(config)#vlan 10
S1(config-vlan10)#switchport interface ethernet 0/0/10
Set the port Ethernet0/0/10 access vlan 10 successfully
S1(config-vlan10)#exit
S1(config)#vlan 20
S1(config-vlan20)#switchport interface ethernet 0/0/11
Set the port Ethernet0/0/11 access vlan 20 successfully
S1(config-vlan20)#exit
S1(config)#interface ethernet 0/0/23-24
S1(config-if-port-range)#switchport mode trunk
Set the port Ethernet0/0/23 mode Trunk successfully
Set the port Ethernet0/0/24 mode Trunk successfully
S1(config-if-port-range)#exit
S1(config)#
```

任务 3：配置多实例生成树，在交换机 S1、S2 上分别将 vlan 10 映射到实例 1 上，将 vlan 20 映射到实例 2 上。

交换机 S1：

```
S1(config)#spanning-tree mst configuration
S1(config-mstp-region)#name mstp
S1(config-mstp-region)#instance 1 vlan 10
S1(config-mstp-region)#instance 2 vlan 20
S1(config-mstp-region)#exit
```

```
S1(config)#spanning-tree

MSTP is starting now, please wait............
MSTP is enabled successfully
S1(config)#
```

交换机 S2：

```
S2(config)#spanning-tree mst configuration
S2(Config-Mstp-Region)#name mstp
S2(Config-Mstp-Region)#instance 1 vlan 10
S2(Config-Mstp-Region)#instance 2 vlan 20
S2(Config-Mstp-Region)#exit
S2(config)#spanning-tree

MSTP is starting now, please wait..............
MSTP is enabled successfully
S2(config)#
```

任务 4：在根交换机中配置端口在不同实例中的优先级，确保不同实例阻塞不同端口。代码如下。

```
S1#show spanning-tree

                -- MSTP Bridge Config Info --

Standard         :   IEEE 802.1s
Bridge MAC       :   00:03:0f:0f:2a:63
Bridge Times :   Max Age 20, Hello Time 2, Forward Delay 15
Force Version:   3

############################ Instance 0 ############################
Self Bridge Id      : 32768 -   00:03:0f:0f:2a:63
Root Id              : this switch
Ext.RootPathCost : 0
Region Root Id      : this switch
Int.RootPathCost : 0
Root Port ID        : 0
…………
```

可以看出 S1 交换机为根交换机，在根交换机上修改 trunk 端口在不同实例中的优先级。代码如下。

```
S1(config)#interface ethernet 0/0/23
S1(config-if-ethernet0/0/23)#spanning-tree mst 1 port-priority 32
```

```
S1(config-if-ethernet0/0/23)#exit
S1(config)#interface ethernet 0/0/24
S1(config-if-ethernet0/0/24)#spanning-tree mst 2 port-priority 32
S1(config-if-ethernet0/0/24)#exit
S1(config)#
```

任务 5：验证多实例生成树。代码如下。

```
S2#show spanning-tree
                    -- MSTP Bridge Config Info --

Standard         :   IEEE 802.1s
Bridge MAC       :   00:03:0f:0f:4a:29
Bridge Times :   Max Age 20, Hello Time 2, Forward Delay 15
Force Version:   3

############################ Instance 0 ############################
Self Bridge Id       : 32768 -    00:03:0f:0f:4a:29
Root Id                 : 32768.00:03:0f:0f:2a:63
Ext.RootPathCost : 0
Region Root Id      : 32768.00:03:0f:0f:2a:63
Int.RootPathCost : 200000
Root Port ID         : 128.23
Current port list in Instance 0:
Ethernet1/10 Ethernet1/11 Ethernet1/23 Ethernet1/24 (Total 4)

   PortName        ID        ExtRPC     IntRPC   State Role       DsgBridge        DsgPort
-------------- ------- --------- --------- --- ---- ------------------
  Ethernet1/10 128.010          0      200000 FWD DSGN 32768.00030f0f4a29 128.010
  Ethernet1/11 128.011          0      200000 FWD DSGN 32768.00030f0f4a29 128.011
  Ethernet1/23 128.023          0           0 FWD ROOT 32768.00030f0f2a63 128.023
  Ethernet1/24 128.024          0           0 BLK ALTR 32768.00030f0f2a63 128.024

############################ Instance 1 ############################
Self Bridge Id     : 32768.00:03:0f:0f:4a:29
Region Root Id     : 32768.00:03:0f:0f:2a:63
Int.RootPathCost : 200000
Root Port ID       : 128.23
Current port list in Instance 1:
Ethernet1/10 Ethernet1/23 Ethernet1/24 (Total 3)
```

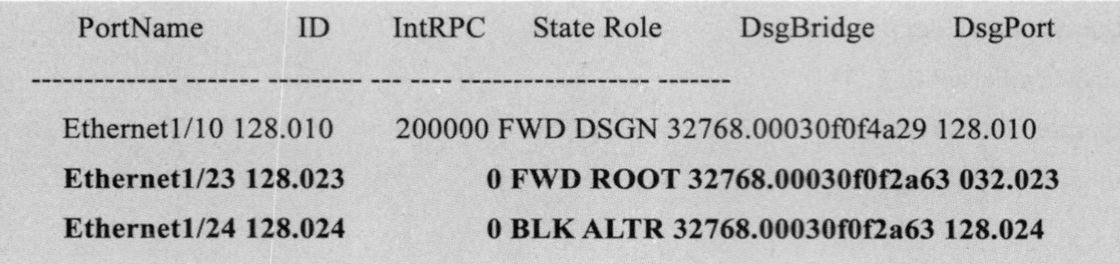

```
   PortName          ID       IntRPC     State Role        DsgBridge          DsgPort
-------------- ------- --------- --- ---- ------------------ -------
  Ethernet1/10 128.010       200000 FWD DSGN 32768.00030f0f4a29 128.010
  Ethernet1/23 128.023                0 FWD ROOT 32768.00030f0f2a63 032.023
  Ethernet1/24 128.024                0 BLK ALTR 32768.00030f0f2a63 128.024
```

```
########################### Instance 2 ###########################
Self Bridge Id      : 32768.00:03:0f:0f:4a:29
Region Root Id      : 32768.00:03:0f:0f:2a:63
Int.RootPathCost : 200000
Root Port ID        : 128.24
Current port list in Instance 2:
Ethernet1/11 Ethernet1/23 Ethernet1/24 (Total 3)

   PortName          ID       IntRPC     State Role        DsgBridge          DsgPort
-------------- ------- --------- --- ---- ------------------ -------
  Ethernet1/11 128.011       200000 FWD DSGN 32768.00030f0f4a29 128.011
  Ethernet1/23 128.023                0 BLK ALTR 32768.00030f0f2a63 128.023
  Ethernet1/24 128.024                0 FWD ROOT 32768.00030f0f2a63 032.024
S2#
```

可以看到在不同实例中 23、24 端口的转发阻塞情况。

实训 5　链 路 聚 合

实 训 目 标

小张：这个链路聚合看上去和生成树有点像。

李工：不，虽然有点形似，但功能完全不同。链路聚合是将几个链路作聚合处理，这几个链路必须是同时连接两个相同的设备，这样，当作了链路聚合之后就可以实现几个链路相加的带宽了。比如，我们可以将 4 个 100Mbit/s 链路使用链路聚合作成一个逻辑链路，这样在全双工条件下就可以达到 800Mbit/s 的带宽，同时可以实现链路负载平衡，提供额外的容错性能。这种方式比较经济，实现也相对容易。

实现链路聚合的条件：

① 端口必须处于相同的 VLAN 之中。

② 端口必须使用相同的传输介质。

③ 端口必须都处于全双工工作模式。

④ 端口必须是相同传输速率的端口。

实训拓扑

实训拓扑图如图 2-9 所示。

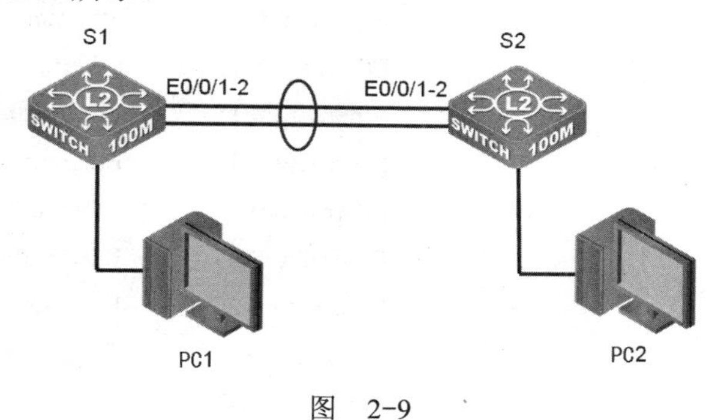

图　2-9

实训任务

任务 1：正确连接并创建端口组。

由于神码交换机默认生成树协议是关闭的，所以在创建链路聚合时只能先连接一条线，避免产生广播风暴，待聚合端口创建成功后再连接其他线缆。

交换机 S1：

```
S1(Config)#port-group 1
S1(Config)#
```

交换机 S2：

```
S2(Config)#port-group 2
S2(Config)#
```

任务 2：静态生成链路聚合通道。

交换机 S1：

```
S1(config)#interface ethernet 0/0/1-2
S1(config-if-port-range)#port-group 1 mode on
S1(config-if-port-range)#
```

交换机 S2：

```
S2(config)#interface ethernet 0/0/1-2
S2(config-if-port-range)#port-group 2 mode on          //注意与交换机 S1 的模式匹配
S2(config-if-port-range)#
```

查看 VLAN 信息：

```
S1#show vlan
VLAN Name         Type       Media      Ports
---- ------------ ---------- --------- ----------------------------------------
1    default      Static     ENET       Ethernet0/0/3        Ethernet0/0/4
```

```
                    Ethernet0/0/5          Ethernet0/0/6
                    Ethernet0/0/7          Ethernet0/0/8
                    Ethernet0/0/9          Ethernet0/0/10
                    Ethernet0/0/11         Ethernet0/0/12
                    Ethernet0/0/13         Ethernet0/0/14
                    Ethernet0/0/15         Ethernet0/0/16
                    Ethernet0/0/17         Ethernet0/0/18
                    Ethernet0/0/19         Ethernet0/0/20
                    Ethernet0/0/21         Ethernet0/0/22
                    Ethernet0/0/23         Ethernet0/0/24
                    Ethernet0/0/25         Ethernet0/0/26
                    Port-Channel1
S1#
```

我们看到多了一个通道 Port-Channel1（在 S2 上显示为 Port-Channel2），表示此时静态聚合链路已创建完成，分别连接 S1 和 S2 交换机的 E0/0/1 和 E0/0/2 端口，使用 PC1 ping PC2。代码如下。

```
C:\Documents and Settings\Administrator>ping 192.168.1.11

Pinging 192.168.1.11 with 32 bytes of data:

Reply from 192.168.1.11: bytes=32 time=2ms TTL=64
Reply from 192.168.1.11: bytes=32 time=1ms TTL=64
Reply from 192.168.1.11: bytes=32 time=1ms TTL=64
Reply from 192.168.1.11: bytes=32 time=1ms TTL=64

Ping statistics for 192.168.1.11:
    Packets: Sent = 4, Received = 4, Lost = 0 (0% loss),
Approximate round trip times in milli-seconds:
    Minimum = 1ms, Maximum = 12ms, Average = 4ms
```

可以正常通信。拔掉聚合链路中的一条链路，再用 PC1 ping PC2。代码如下。

```
C:\Documents and Settings\Administrator>ping 192.168.1.11

Pinging 192.168.1.11 with 32 bytes of data:

Reply from 192.168.1.11: bytes=32 time=2ms TTL=64
Reply from 192.168.1.11: bytes=32 time=2ms TTL=64
Reply from 192.168.1.11: bytes=32 time=2ms TTL=64
Reply from 192.168.1.11: bytes=32 time=1ms TTL=64
```

```
Ping statistics for 192.168.1.11:
    Packets: Sent = 4, Received = 4, Lost = 0 (0% loss),
Approximate round trip times in milli-seconds:
    Minimum = 1ms, Maximum = 12ms, Average = 4ms
```

依然通信正常，静态链路聚合创建完成。

任务 3：LACP 动态生成链路聚合通道。

启动 LACP 的端口可以有两种工作模式：passive，和 active。

Passive：被动模式，该模式下端口不会主动发送 LACPDU 报文，在接收到对端发送的 LACP 报文后，该端口进入协议计算状态。

Active：主动模式，该模式下端口会主动向对端发送 LACPDU 报文，进行 LACP 的计算。

交换机 S1：

```
S1(Config)#port-group 1
S1(config)#interface ethernet 0/0/1-2
S1(config-if-port-range)#port-group 1 mode active
S1(config-if-port-range)#
```

交换机 S2：

```
S2(Config)#port-group 2
S2(config)#interface ethernet 0/0/1-2
S2(config-if-port-range)#port-group 2 mode passive
S2(config-if-port-range)#
```

验证 S1：

```
S1#show interface port-channel 1
Interface brief:
  Port-Channel1 is up, line protocol is up
  Port-Channel1 is layer 2 port, alias name is (null), index is 29
  Port-Channel1 is LAG port, member is :
    Ethernet0/0/1      Ethernet0/0/3
  Hardware is EtherChannel, address is 00-03-0f-1a-ec-3d
```

我们看到聚合组 Port-Channel1 已经 up，聚合端口有 E0/0/1 和 E0/0/3 两个端口。

验证 S2：

```
S2# show interface port-channel 2
Interface brief:
  Port-Channel2 is up, line protocol is up
  Port-Channel2 is layer 2 port, alias name is (null), index is 30
  Port-Channel2 is LAG port, member is :
    Ethernet0/0/1      Ethernet0/0/3
  Hardware is EtherChannel, address is 00-03-0f-1a-ec-49
```

S2 上的聚合组 Port-Channel2 也已经 up。

PC1 ping PC2：

```
C:\Documents and Settings\Administrator>ping 192.168.1.11

Pinging 192.168.1.11 with 32 bytes of data:

Reply from 192.168.1.11: bytes=32 time=2ms TTL=64
Reply from 192.168.1.11: bytes=32 time=1ms TTL=64
Reply from 192.168.1.11: bytes=32 time=1ms TTL=64
Reply from 192.168.1.11: bytes=32 time=1ms TTL=64

Ping statistics for 192.168.1.11:
    Packets: Sent = 4, Received = 4, Lost = 0 (0% loss),
Approximate round trip times in milli-seconds:
    Minimum = 1ms, Maximum = 12ms, Average = 4ms
```

动态链路聚合完成。

实训6 端口安全

实训目标

李工：有时为了安全和管理上的需要，需将设备或终端的 MAC 地址与交换机端口进行绑定，端口只允许已绑定 MAC 的数据流的转发。即，MAC 地址与端口绑定后，该 MAC 地址的数据流只能从绑定端口进入，其他没有与端口绑定的 MAC 地址的数据流不可以从该端口进入。

实训拓扑

实训拓扑图如图 2-10 所示。

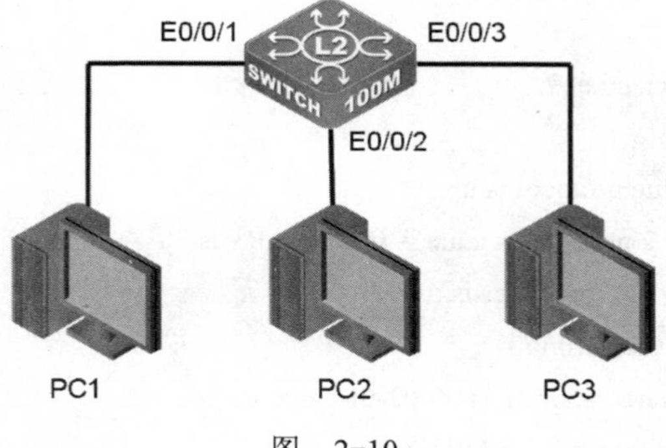

图　2-10

实训任务

任务 1：规划 PC 的 IP 地址与端口。见表 2-6。

表　2-6

PC 机	IP 地址	掩码	MAC 地址	端口
PC1	192.168.1.10	255.255.255.0	00-0b-4c-9e-2a-1c	E0/0/1
PC2	192.168.1.11	255.255.255.0	00-0b-2e-5d-62-4e	E0/0/2
PC3	192.168.1.12	255.255.255.0	00-0b-cd-4a-97-08	E0/0/3

任务 2：静态绑定 MAC 地址。

E0/0/1 端口绑定 PC1 的 MAC 地址。代码如下。

```
S1(config-if-ethernet0/0/1)#switchport port-security
S1(config-if-ethernet0/0/1)#switchport port-security mac-address 00-0b-4c-9e-2a-1c
```

PC1 分别 Ping PC2 和 PC3 可以 Ping 通。代码如下。

```
C:\Documents and Settings\Administrator>ping 192.168.1.12

Pinging 192.168.1.12 with 32 bytes of data:

Reply from 192.168.1.12: bytes=32 time<1ms TTL=64
Reply from 192.168.1.12: bytes=32 time<1ms TTL=64
Reply from 192.168.1.12: bytes=32 time<1ms TTL=64
Reply from 192.168.1.12: bytes=32 time<1ms TTL=64

Ping statistics for 192.168.1.12:
    Packets: Sent = 4, Received = 4, Lost = 0 (0% loss),
Approximate round trip times in milli-seconds:
    Minimum = 0ms, Maximum = 0ms, Average = 0ms
```

把 PC1 更换到 E0/0/4 端口，再次 Ping PC2 和 PC3，已无法 Ping 通。代码如下。

```
C:\Documents and Settings\Administrator>ping 192.168.1.12

Pinging 192.168.1.12 with 32 bytes of data:

Request timed out.
Request timed out.
Request timed out.
Request timed out.

Ping statistics for 192.168.1.12:
    Packets: Sent = 4, Received = 0, Lost = 4 (100% loss),
```

可见，PC1 只能通过其绑定的 E0/0/1 端口与网络进行通信。

小张：如果把其他 PC 换到 E0/0/1 口是什么情况呢？

李工：你试下。

小张把 PC2 更换到 E0/0/1 口，分别 Ping E0/0/3 口的 PC3 和 E0/0/4 口的 PC1。代码如下。

```
C:\Documents and Settings\Administrator>ping 192.168.1.12

Pinging 192.168.1.12 with 32 bytes of data:

Reply from 192.168.1.12: bytes=32 time<1ms TTL=64
Reply from 192.168.1.12: bytes=32 time<1ms TTL=64
Reply from 192.168.1.12: bytes=32 time<1ms TTL=64
Reply from 192.168.1.12: bytes=32 time<1ms TTL=64

Ping statistics for 192.168.1.12:
    Packets: Sent = 4, Received = 4, Lost = 0 (0% loss),
Approximate round trip times in milli-seconds:
    Minimum = 0ms, Maximum = 0ms, Average = 0ms

C:\Documents and Settings\Administrator>ping 192.168.1.10

Pinging 192.168.1.10 with 32 bytes of data:

Request timed out.
Request timed out.
Request timed out.
Request timed out.

Ping statistics for 192.168.1.10:
    Packets: Sent = 4, Received = 0, Lost = 4 (100% loss),
```

小张：为什么 PC2 还能 Ping 通 PC3 呢？

李工：端口 E0/0/1 绑定了 PC1，依然可以与其他 MAC 地址通信，但 PC1 只能通过 E0/0/1 端口进行通信。如果我们在 E0/0/1 口加入指令：

```
S1(config-if-ethernet0/0/1)#switchport port-security lock
```

再用 PC2 Ping PC3：

```
C:\Documents and Settings\Administrator>ping 192.168.1.12

Pinging 192.168.1.12 with 32 bytes of data:

Request timed out.
```

```
Request timed out.
Request timed out.
Request timed out.

Ping statistics for 192.168.1.12:
    Packets: Sent = 4, Received = 0, Lost = 4 (100% loss),
```

就无法 Ping 通了。

把 PC1 换回 E0/0/1 端口，Ping PC3：

```
C:\Documents and Settings\Administrator>ping 192.168.1.12

Pinging 192.168.1.12 with 32 bytes of data:

Reply from 192.168.1.12: bytes=32 time<1ms TTL=64
Reply from 192.168.1.12: bytes=32 time<1ms TTL=64
Reply from 192.168.1.12: bytes=32 time<1ms TTL=64
Reply from 192.168.1.12: bytes=32 time<1ms TTL=64

Ping statistics for 192.168.1.12:
    Packets: Sent = 4, Received = 4, Lost = 0 (0% loss),
Approximate round trip times in milli-seconds:
    Minimum = 0ms, Maximum = 0ms, Average = 0ms
```

我们看到，此时 E0/0/1 端口只能接受绑定的 MAC 地址，而不能接受其他 MAC 地址通过此端口进行通信。

小张：那是不是一个端口只能静态绑定一个 MAC 地址呢？

李工：当然不是。默认情况下，交换机允许绑定一个 MAC 地址，但可以让一个端口绑定多个 MAC 地址。

任务 3：静态绑定多个 MAC 地址。

取消端口 E0/0/1 锁定，更改最大 MAC 地址数，然后绑定 PC3 的 MAC 地址。代码如下。

```
S1(config-if-ethernet0/0/1)#no switchport port-security lock
S1(config-if-ethernet0/0/1)#switchport port-security maximum 2
S1(config-if-ethernet0/0/1)#switchport port-security mac-address 00-0b-cd-4a-97-08
S1(config-if-ethernet0/0/1)#switchport port-security lock
```

查看绑定的 MAC 地址情况。代码如下。

```
S1#show port-security address
Security Mac Address Table
---------------------------------------------------------------------------
Vlan     Mac Address              Type                 Ports
 1      00-0b-4c-9e-2a-1c       SecurityConfigured   Ethernet0/0/1
 1      00-0b-cd-4a-97-08       SecurityConfigured   Ethernet0/0/1
```

```
---------------------------------------------------------------------------
Total Addresses in System :2
Max Addresses limit in System :128


S1#
```

可以看出，E0/0/1 端口已绑定了 2 个 MAC 地址，最多可以绑定 128 个 MAC 地址。

把 PC3 插在 E0/0/1 端口，PC1 插在 E0/0/3 端口，PC2 端口不变，用 PC3 Ping PC2。代码如下。

```
C:\Documents and Settings\Administrator>ping 192.168.1.11

Pinging 192.168.1.12 with 32 bytes of data:

Reply from 192.168.1.12: bytes=32 time=1ms TTL=64
Reply from 192.168.1.12: bytes=32 time<1ms TTL=64
Reply from 192.168.1.12: bytes=32 time<1ms TTL=64
Reply from 192.168.1.12: bytes=32 time<1ms TTL=64

Ping statistics for 192.168.1.12:
    Packets: Sent = 4, Received = 4, Lost = 0 (0% loss),
Approximate round trip times in milli-seconds:
    Minimum = 0ms, Maximum = 0ms, Average = 0ms
```

可以 Ping 通。PC3 如果现在 Ping PC1。代码如下。

```
C:\Documents and Settings\Administrator>ping 192.168.1.10

Pinging 192.168.1.10 with 32 bytes of data:

Request timed out.
Request timed out.
Request timed out.
Request timed out.

Ping statistics for 192.168.1.10:
    Packets: Sent = 4, Received = 0, Lost = 4 (100% loss),
```

这是因为 PC1 已绑定在 E0/0/1 端口，不能在其他端口通信。

任务 4：动态绑定 MAC 地址。

通常情况下，并不需要特别指定是哪台设备必须绑定在哪个端口。这时，可以使用动态绑定方法，让交换机端口自动绑定连接它的设备。

还原 PC 与交换机端口的连接，重启交换机还原到出厂设置。代码如下。

```
S1#set default
```

```
Are you sure? [Y/N] = y
S1#reload
Process with reboot? [Y/N] y
```

设置 E0/0/1 端口自动绑定连接它的 MAC 地址。代码如下。

```
S1(config-if-ethernet0/0/1)#switchport port-security
S1(config-if-ethernet0/0/1)#switchport port-security lock
S1(config-if-ethernet0/0/1)#switchport port-security convert
1 dynamic mac have been converted to security mac on interface Ethernet0/0/1
S1(config-if-ethernet0/0/1)#
```

查看端口绑定地址。代码如下。

```
S1#show port-security address
Security Mac Address Table
---------------------------------------------------------------------------
Vlan    Mac Address              Type                  Ports
 1     00-0b-4c-9e-2a-1c        SecurityConfigured    Ethernet0/0/1
---------------------------------------------------------------------------
Total Addresses in System :1
Max Addresses limit in System :128


S1#
```

任务 5：绑定 MAC-IP 地址。

MAC-IP 地址绑定可以将端口绑定为指定的设备和 IP 地址，更换设备或 IP 地址均无法通过此端口访问。代码如下。

```
S1(config)#am enable
S1(config)#interface ethernet1/1
S1(config-if-ethernet1/1)#am port
S1(config-if-ethernet1/1)#am mac-ip-pool 00-0b-4c-9e-2a-1c 192.168.1.10
```

可将端口 E1/1 绑定为 PC1，此时 PC1 可通过该端口访问 PC2 或 PC3。若更改 PC1 的 IP 地址，或将其他设备（如 PC2）连接 E1/1 端口则无法进行访问。

需要注意的是，PC1 仍可通过未绑定地址的端口访问其他设备。

任务 6：绑定 IP 地址区间。

将端口绑定一个 IP 地址区间，则此区间的 IP 地址设备可通过该端口进行访问，不在此区间的 IP 地址无法通过该端口访问网络。

将端口 E1/2 绑定 192.168.1.10 和 192.168.1.11 两个 IP 地址。代码如下。

```
S1(config)#am enable
S1(config)#interface ethernet 1/2
S1(config-if-ethernet1/2)#am port
S1(config-if-ethernet1/2)#am ip-pool 192.168.1.10 2     //绑定 192.168.1.10 开始的 2 个地址
```

除 192.168.1.10 和 192.168.1.11 外的其他 IP 地址均无法通过此端口通信。

第3章　交换机路由应用实践

实训1　单臂路由实现 VLAN 间互访

实训目标

李工：划分 VLAN 可以隔离广播域，有效提高网络性能。但 VLAN 之间无法通信，这并不是我们的目的。既能隔离广播域，防止广播风暴的发生，又能实现 VLAN 之间的通信，就需要网络层的设备来支持。常见的网络层设备是路由器，可以通过路由器以单臂路由方式来实现 VLAN 之间的通信。

实训拓扑

实训拓扑图如图 3-1 所示。

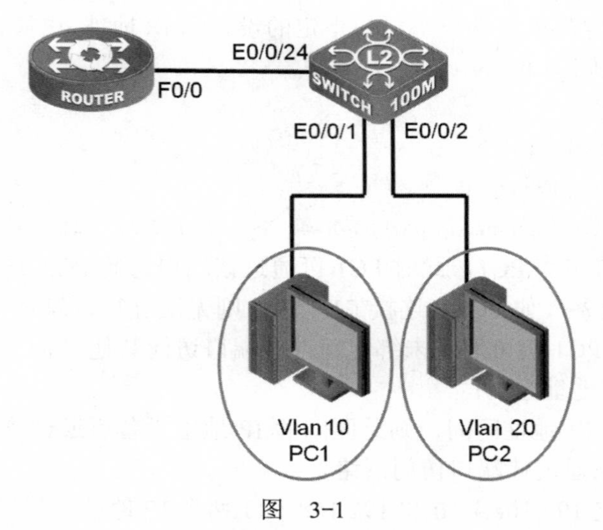

图　3-1

实训任务

任务 1：规划交换机端口、VLAN 及 PC IP 地址。见表 3-1。

表　3-1

交换机端口	所属 VLAN	连 接 设 备	IP　地　址	网　　关
E1/1	VLAN10	PC1	192.168.1.10	192.168.1.1
E1/2	VLAN20	PC2	192.168.2.10	192.168.2.1
E1/24	Trunk 口，连接路由器 F0/0 以太网口			

任务 2：配置交换机。代码如下。

```
S1(config)#vlan 10
S1(config-vlan10)#switchport interface ethernet 1/1
Set the port Ethernet1/1 access vlan 10 successfully
S1(config-vlan10)#exit
S1(config)#vlan 20
S1(config-vlan20)#switchport interface ethernet 1/2
Set the port Ethernet1/2 access vlan 20 successfully
S1(config-vlan20)#exit
S1(config)#interface ethernet 1/24
S1(config-if-ethernet1/24)#switchport mode trunk
Set the port Ethernet1/24 mode Trunk successfully
S1(config-if-ethernet1/24)#switchport trunk allowed vlan all
S1(config-if-ethernet1/24)#exit
S1(config)#
```

任务 3：配置路由器。代码如下。

```
Router>enable
Router#config
Router_config#
Router_config#interface f0/0.1                          //配置子接口
Router_config_f0/0.1#encapsulation dot1Q 10             //封装 802.1q
Router_config_f0/0.1#ip address 192.168.1.1 255.255.255.0       //配置 vlan10 的网关地址
Router_config_f0/0.1#exit
Router_config#interface f0/0.2
Router_config_f0/0.2#encapsulation dot1Q 20
Router_config_f0/0.2#ip address 192.168.2.1 255.255.255.0
Router_config_f0/0.2#exit
Router_config#
```

单臂路由配置完成，用 PC1 Ping PC2，可以看到通信正常。代码如下。

```
C:\Documents and Settings\Administrator>ping 192.168.2.10

Pinging 192.168.2.10 with 32 bytes of data:

Reply from 192.168.2.10: bytes=32 time=1ms TTL=63
```

```
Reply from 192.168.2.10: bytes=32 time<1ms TTL=63
Reply from 192.168.2.10: bytes=32 time<1ms TTL=63
Reply from 192.168.2.10: bytes=32 time<1ms TTL=63

Ping statistics for 192.168.2.10:
    Packets: Sent = 4, Received = 4, Lost = 0 (0% loss),
Approximate round trip times in milli-seconds:
    Minimum = 0ms, Maximum = 1ms, Average = 0ms
```

实训 2　三层交换机实现 VLAN 间互访

实训目标

李工：单臂路由方式受带宽限制，在大型网络中对网络性能影响较大。这时通常采用三层交换机。三层交换机在二层交换机基础上增加了网络层的功能，交换机以太网口较多，不受带宽限制，可用于较大型网络的 VLAN 间通信。

实训拓扑

实训拓扑图如图 3-2 所示。

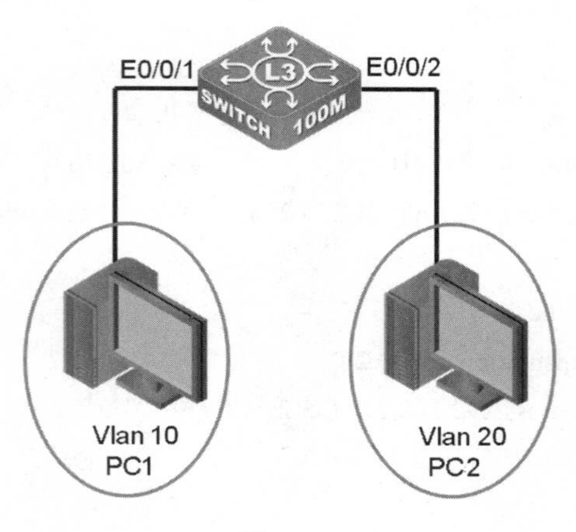

图　3-2

实训任务

任务 1：规划交换机端口、VLAN 及 PC IP 地址。见表 3-2。

表　3-2

VLAN	端　口	VLAN 接口 IP	连　接　PC	IP　地　址	网　关
10	E1/1	192.168.1.1	PC1	192.168.1.10	192.168.1.1
20	E1/2	192.168.2.1	PC2	192.168.2.10	192.168.2.1

任务 2：配置三层交换机。

划分 VLAN 10 及 VLAN 20，分别加入端口 E1/1 和 E1/2，配置各 VLAN 的接口地址。代码如下。

```
S1(config)#vlan 10
S1(Config-Vlan10)#switchport interface ethernet 1/1
Set the port Ethernet1/1 access vlan 10 successfully
S1(Config-Vlan10)#exit
S1(config)#vlan 20
S1(Config-Vlan20)#switchport interface ethernet 1/2
Set the port Ethernet1/2 access vlan 20 successfully
S1(Config-Vlan20)#exit
S1(config)#interface vlan 10
S1(Config-if-Vlan10)#ip address 192.168.1.1 255.255.255.0
S1(Config-if-Vlan10)#exit
S1(config)#interface vlan 20
S1(Config-if-Vlan20)#ip address 192.168.2.1 255.255.255.0
S1(Config-if-Vlan20)#exit
S1(config)#
```

查看交换机路由表。代码如下。

```
S1#show ip router
Codes: K - kernel, C - connected, S - static, R - RIP, B - BGP
       O - OSPF, IA - OSPF inter area
       N1 - OSPF NSSA external type 1, N2 - OSPF NSSA external type 2
       E1 - OSPF external type 1, E2 - OSPF external type 2
       i - IS-IS, L1 - IS-IS level-1, L2 - IS-IS level-2, ia - IS-IS inter area
       * - candidate default

C        127.0.0.0/8 is directly connected, Loopback
C        192.168.1.0/24 is directly connected, Vlan10
C        192.168.2.0/24 is directly connected, Vlan20
Total routes are : 3 item(s)
S1#
```

可以看到路由表中已包括直连路由 VLAN 10 和 VLAN 20 网段。

PC1 Ping PC2，代码如下。

```
C:\Documents and Settings\Administrator>ping 192.168.2.10
```

```
Pinging 192.168.2.10 with 32 bytes of data:

Reply from 192.168.2.10: bytes=32 time=2ms TTL=63
Reply from 192.168.2.10: bytes=32 time<1ms TTL=63
Reply from 192.168.2.10: bytes=32 time<1ms TTL=63
Reply from 192.168.2.10: bytes=32 time<1ms TTL=63

Ping statistics for 192.168.2.10:
    Packets: Sent = 4, Received = 4, Lost = 0 (0% loss),
Approximate round trip times in milli-seconds:
    Minimum = 0ms, Maximum = 0ms, Average = 0ms
```

可看到 PC1 与 PC2 通信正常。

实训 3　动态路由协议 RIP

实训目标

李工：当两台以上三层交换机级联时，为了使各交换机上连接的不同网段之间能够互相通信，这时就需要配置动态路由协议。常见的动态路由协议有 RIP、OSPF 等。

RIP（Routing Information Protocol，路由信息协议）基于距离矢量算法，它使用“跳数”，即 metric 来衡量到达目标地址的路由距离。这种协议只关心自己周围的世界，只与自己相邻的网络设备交换信息，范围限制在 15 跳之内，再远，它就不关心了。

实训拓扑

实训拓扑图如图 3-3 所示。

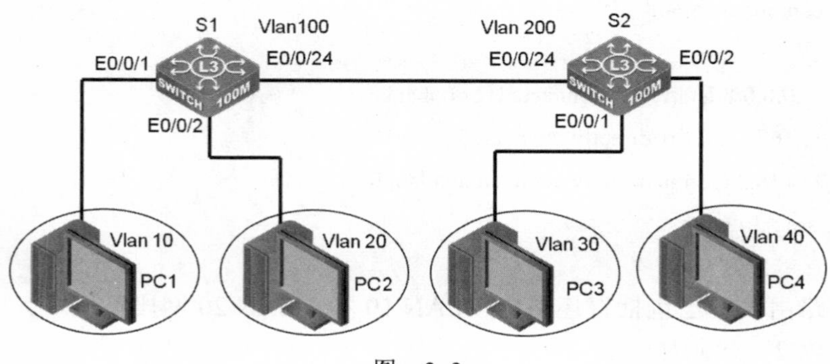

图　3-3

实训任务

任务 1：环境规划。

1）在交换机 A 和交换机 B 上分别划分基于端口的 VLAN，见表 3-3。

表 3-3

交换机	VLAN	端口成员
交换机 S1	10	E1/1
	20	E1/2
	100	E1/24
交换机 S2	30	E1/1
	40	E1/2
	200	E1/24

2）交换机 S1 和 S2 通过的 24 口级联。

3）配置交换机 S1 和 S2 各 VLAN 接口的 IP 地址，见表 3-4。

表 3-4

VLAN 10	VLAN 20	VLAN 30	VLAN 40	VLAN 100	VLAN 200
192.168.10.1	192.168.20.1	192.168.30.1	192.168.40.1	192.168.100.1	192.168.100.2

4）PC1-PC4 的网络设置，见表 3-5。

表 3-5

设备	IP 地址	网关	掩码
PC1	192.168.10.10	192.168.10.1	255.255.255.0
PC2	192.168.20.10	192.168.20.1	255.255.255.0
PC3	192.168.30.10	192.168.30.1	255.255.255.0
PC4	192.168.40.10	192.168.40.1	255.255.255.0

任务 2：配置交换机 S1。代码如下。

```
S1(config)#vlan 10
S1(Config-Vlan10)#switchport interface ethernet 1/1
Set the port Ethernet1/1 access vlan 10 successfully
S1(Config-Vlan10)#exit
S1(config)#vlan 20
S1(Config-Vlan20)#switchport interface ethernet 1/2
Set the port Ethernet1/2 access vlan 20 successfully
S1(Config-Vlan20)#exit
S1(config)#vlan 100
S1(Config-Vlan100)#switchport interface ethernet 1/24
Set the port Ethernet1/24 access vlan 100 successfully
S1(Config-Vlan20)#exit
S1(config)#interface vlan 10
S1(Config-if-Vlan10)#ip address 192.168.10.1 255.255.255.0
```

```
S1(Config-if-Vlan10)#exit
S1(config)#interface vlan 20
S1(Config-if-Vlan20)#ip address 192.168.20.1 255.255.255.0
S1(Config-if-Vlan20)#exit
S1(config)#interface vlan 100
S1(Config-if-Vlan20)#ip address 192.168.100.1 255.255.255.0
S1(Config-if-Vlan20)#exit
S1(config)#
```

任务 3：配置交换机 S2。代码如下。

```
S2(config)#vlan 30
S2(config-vlan30)#switchport interface ethernet 1/0/1
Set the port Ethernet1/0/1 access vlan 30 successfully
S2(config-vlan30)#exit
S2(config)#vlan 40
S2(config-vlan40)#switchport interface ethernet 1/0/2
Set the port Ethernet1/0/2 access vlan 40 successfully
S2(config-vlan40)#exit
S2(config)#vlan 200
S2(config-vlan200)#switchport interface ethernet 1/0/24
Set the port Ethernet1/0/24 access vlan 200 successfully
S2(config-vlan200)#exit
S2(config)#interface vlan 30
S2(config-if-vlan30)#ip address 192.168.30.1 255.255.255.0
S2(config-if-vlan30)#exit
S2(config)#interface vlan 40
S2(config-if-vlan40)#ip address 192.168.40.1 255.255.255.0
S2(config-if-vlan40)#exit
S2(config)#interface vlan 200
S2(config-if-vlan200)#ip address 192.168.100.2 255.255.255.0
S2(config-if-vlan200)#exit
S2(config)#
```

任务 4：验证互通性。见表 3-6。

表 3-6

源 PC	端 口	目 的 PC	端 口	结 果
PC1	E1/1	PC2	S1：E1/2	通
PC1	E1/1	VLAN100	S1：E1/24	通
PC1	E1/1	PC3	S2：E1/0/2	不通
PC1	E1/1	VLAN200	S2：E1/0/24	不通

查看 S1 路由表：

```
S1#show ip router
Codes: K - kernel, C - connected, S - static, R - RIP, B - BGP
       O - OSPF, IA - OSPF inter area
       N1 - OSPF NSSA external type 1, N2 - OSPF NSSA external type 2
       E1 - OSPF external type 1, E2 - OSPF external type 2
       i - IS-IS, L1 - IS-IS level-1, L2 - IS-IS level-2, ia - IS-IS inter area
       * - candidate default

C       127.0.0.0/8 is directly connected, Loopback
C       192.168.10.0/24 is directly connected, Vlan10
C       192.168.20.0/24 is directly connected, Vlan20
C       192.168.100.0/24 is directly connected, Vlan100
Total routes are : 4 item(s)
S1#
```

可以看到 S1 路由表中只有直连路由信息，即 192.168.10.0、192.168.20.0、192.168.100.0 网段，并没有 S2 交换机上的路由信息，因此 PC1 无法 Ping 通 S2 交换机上的网段。

任务 5：分别在 S1 和 S2 上添加各自直连网段的动态路由协议 RIP。代码如下。

```
S1(config)#router rip
S1(config-router)#network vlan 10
S1(config-router)#network vlan 20
S1(config-router)#network vlan 100
S1(config-router)#exit
S1(config)#

S2(config)#router rip
S2(config-router)#network vlan 30
S2(config-router)#network vlan 40
S2(config-router)#network vlan 200
S2(config-router)#exit
S2(config)#
```

查看 S1 路由表：

```
S1#show ip router
Codes: K - kernel, C - connected, S - static, R - RIP, B - BGP
       O - OSPF, IA - OSPF inter area
       N1 - OSPF NSSA external type 1, N2 - OSPF NSSA external type 2
       E1 - OSPF external type 1, E2 - OSPF external type 2
       i - IS-IS, L1 - IS-IS level-1, L2 - IS-IS level-2, ia - IS-IS inter area
       * - candidate default
```

```
C         127.0.0.0/8 is directly connected, Loopback
C         192.168.10.0/24 is directly connected, Vlan10
C         192.168.20.0/24 is directly connected, Vlan20
R         192.168.30.0/24 [120/2] via 192.168.100.2, Vlan100, 00:03:05
R         192.168.40.0/24 [120/2] via 192.168.100.2, Vlan100, 00:03:05
C         192.168.100.0/24 is directly connected, Vlan100
Total routes are : 6 item(s)
S1#
```

验证互通性，见表 3-7。

表 3-7

源 PC	端 口	目 的 PC	端 口	结 果
PC1	E1/1	PC2	S1：E1/2	通
PC1	E1/1	VLAN 100	S1：E1/24	通
PC1	E1/1	PC3	S2：E1/0/2	通
PC1	E1/1	VLAN 200	S2：E1/0/24	通

实训 4 动态路由协议 OSPF

实训目标

李工：相对于 RIP，OSPF（Open Shortest Path First，开放式最短路径优先）协议是一种典型的链路状态（Link-state）的路由协议，一般用于同一个路由域内。OSPF 将链路状态广播数据（Link State Advertisement，LSA）传送给在某一区域内的所有路由器，这一点与 RIP 不同。因此，OSPF 协议能够计算出最优路径，收敛速度快，常应用于大型网络。

实训拓扑

实训拓扑图如图 3-4 所示。

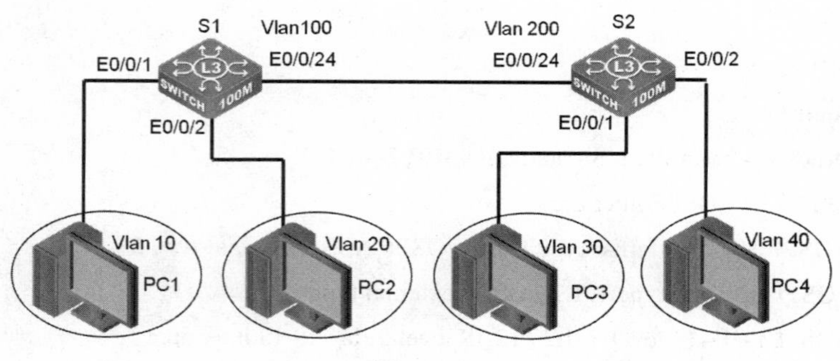

图 3-4

实训任务

任务 1：环境规划。

1）在交换机 A 和交换机 B 上分别划分基于端口的 VLAN，见表 3-8。

表　3-8

交换机	VLAN	端口成员
交换机 S1	10	E1/1
	20	E1/2
	100	E1/24
交换机 S2	30	E1/1
	40	E1/2
	200	E1/24

2）交换机 S1 和 S2 通过的 24 口级联。

3）配置交换机 S1 和 S2 各 VLAN 接口的 IP 地址，见表 3-9。

表　3-9

VLAN 10	VLAN 20	VLAN 30	VLAN 40	VLAN 100	VLAN 200
192.168.10.1	192.168.20.1	192.168.30.1	192.168.40.1	192.168.100.1	192.168.100.2

4）PC1～PC4 的网络设置，见表 3-10。

表　3-10

设备	IP 地址	网关	掩码
PC1	192.168.10.10	192.168.10.1	255.255.255.0
PC2	192.168.20.10	192.168.20.1	255.255.255.0
PC3	192.168.30.10	192.168.30.1	255.255.255.0
PC4	192.168.40.10	192.168.40.1	255.255.255.0

任务 2：配置交换机 S1。代码如下。

```
S1(config)#vlan 10
S1(Config-Vlan10)#switchport interface ethernet 1/1
Set the port Ethernet1/1 access vlan 10 successfully
S1(Config-Vlan10)#exit
S1(config)#vlan 20
S1(Config-Vlan20)#switchport interface ethernet 1/2
Set the port Ethernet1/2 access vlan 20 successfully
S1(Config-Vlan20)#exit
S1(config)#vlan 100
S1(Config-Vlan100)#switchport interface ethernet 1/24
Set the port Ethernet1/24 access vlan 100 successfully
S1(Config-Vlan20)#exit
S1(config)#interface vlan 10
```

```
S1(Config-if-Vlan10)#ip address 192.168.10.1 255.255.255.0
S1(Config-if-Vlan10)#exit
S1(config)#interface vlan 20
S1(Config-if-Vlan20)#ip address 192.168.20.1 255.255.255.0
S1(Config-if-Vlan20)#exit
S1(config)#interface vlan 100
S1(Config-if-Vlan20)#ip address 192.168.100.1 255.255.255.0
S1(Config-if-Vlan20)#exit
S1(config)#
```

任务 3：配置交换机 S2。代码如下。

```
S2(config)#vlan 30
S2(config-vlan30)#switchport interface ethernet 1/0/1
Set the port Ethernet1/0/1 access vlan 30 successfully
S2(config-vlan30)#exit
S2(config)#vlan 40
S2(config-vlan40)#switchport interface ethernet 1/0/2
Set the port Ethernet1/0/2 access vlan 40 successfully
S2(config-vlan40)#exit
S2(config)#vlan 200
S2(config-vlan200)#switchport interface ethernet 1/0/24
Set the port Ethernet1/0/24 access vlan 200 successfully
S2(config-vlan200)#exit
S2(config)#interface vlan 30
S2(config-if-vlan30)#ip address 192.168.30.1 255.255.255.0
S2(config-if-vlan30)#exit
S2(config)#interface vlan 40
S2(config-if-vlan40)#ip address 192.168.40.1 255.255.255.0
S2(config-if-vlan40)#exit
S2(config)#interface vlan 200
S2(config-if-vlan200)#ip address 192.168.100.2 255.255.255.0
S2(config-if-vlan200)#exit
S2(config)#
```

任务 4：验证互通性。见表 3-11。

表 3-11

源 PC	端　口	目 的 PC	端　口	结　果
PC1	E1/1	PC2	S1：E1/2	通
PC1	E1/1	VLAN 100	S1：E1/24	通
PC1	E1/1	PC3	S2：E1/0/2	不通
PC1	E1/1	VLAN 200	S2：E1/0/24	不通

查看 S1 路由表：

```
S1#show ip router
Codes: K - kernel, C - connected, S - static, R - RIP, B - BGP
       O - OSPF, IA - OSPF inter area
       N1 - OSPF NSSA external type 1, N2 - OSPF NSSA external type 2
       E1 - OSPF external type 1, E2 - OSPF external type 2
       i - IS-IS, L1 - IS-IS level-1, L2 - IS-IS level-2, ia - IS-IS inter area
       * - candidate default

C      127.0.0.0/8 is directly connected, Loopback
C      192.168.10.0/24 is directly connected, Vlan10
C      192.168.20.0/24 is directly connected, Vlan20
C      192.168.100.0/24 is directly connected, Vlan100
Total routes are : 4 item(s)
S1#
```

可以看到 S1 路由表中只有直连路由信息，即 192.168.10.0、192.168.20.0、192.168.100.0 网段，并没有 S2 交换机上的路由信息，因此 PC1 无法 Ping 通 S2 交换机上的网段。

任务 5：分别在 S1 和 S2 上启动 OSPF 协议，并将对应的直连网段配置到 OSPF 进程中。代码如下。

```
S1(config)#router ospf
S1(config-router)#network 192.168.10.0/24 area 0
S1(config-router)#network 192.168.20.0/24 area 0
S1(config-router)#network 192.168.100.0/24 area 0
S1(config-router)#exit
S1(config)#

S2(config)#router ospf
S2(config-router)#network 192.168.30.0/24 area 0
S2(config-router)#network 192.168.40.0/24 area 0
S2(config-router)#network 192.168.100.0/24 area 0
S2(config-router)#exit
S2(config)#
```

查看 S1 路由表：

```
S1#show ip router
Codes: K - kernel, C - connected, S - static, R - RIP, B - BGP
       O - OSPF, IA - OSPF inter area
       N1 - OSPF NSSA external type 1, N2 - OSPF NSSA external type 2
       E1 - OSPF external type 1, E2 - OSPF external type 2
       i - IS-IS, L1 - IS-IS level-1, L2 - IS-IS level-2, ia - IS-IS inter area
```

```
        * - candidate default

C       127.0.0.0/8 is directly connected, Loopback    tag:0
C       192.168.10.0/24 is directly connected, Vlan10    tag:0
C       192.168.20.0/24 is directly connected, Vlan20    tag:0
O       192.168.30.0/24 [110/11] via 192.168.100.1, Vlan100, 00:01:10    tag:0
O       192.168.40.0/24 [110/11] via 192.168.100.1, Vlan100, 00:01:10    tag:0
C       192.168.100.0/24 is directly connected, Vlan100    tag:0
Total routes are : 6 item(s)
S1#
```

验证互通性，见表 3-12。

表 3-12

源 PC	端 口	目 的 PC	端 口	结 果
PC1	E1/1	PC2	S1：E1/2	通
PC1	E1/1	VLAN 100	S1：E1/24	通
PC1	E1/1	PC3	S2：E1/0/2	通
PC1	E1/1	VLAN 200	S2：E1/0/24	通

第 4 章　交换机高级应用实践

实训 1　标准访问控制列表

实训目标

李工：访问控制列表（Access Control Lists，ACL）是一种数据包过滤机制，通过允许或拒绝特定的数据包进出网络，使交换机可以对网络访问进行控制，有效保证网络的安全运行。用户可以基于报文中的特定信息制定一组规则（Rule），每条规则都描述了对匹配一定信息的数据包所采取的动作：允许（Permit）或拒绝（Deny）通过。用户可以把这些规则应用到特定交换机端口的入口或出口方向，这样特定端口上特定方向的数据流就必须依照指定的 ACL 规则进出交换机。通过 ACL，可以限制某个 IP 地址的 PC 或者某些网段的 PC 的上网活动。

实训拓扑

实训拓扑图如图 4-1 所示。

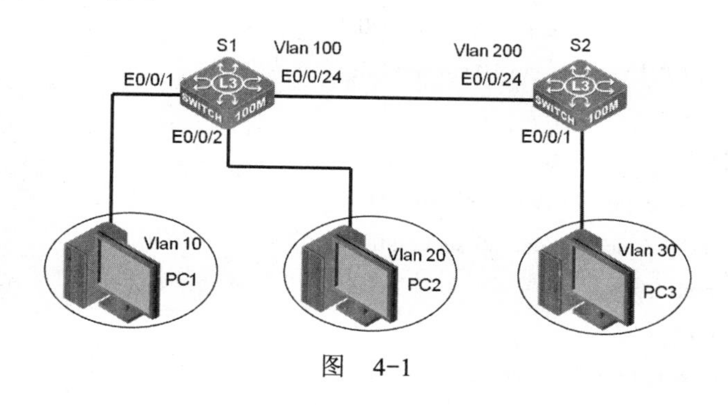

图　4-1

实训任务

任务 1：环境规划。

1）在交换机 A 和交换机 B 上分别划分基于端口的 VLAN，见表 4-1。

表 4-1

交换机	VLAN	端口成员
交换机 S1	10	E0/0/1
	20	E0/0/2
	100	E0/0/24
交换机 S2	30	E0/0/1
	200	E0/0/24

2）交换机 S1 和 S2 通过的 24 口级联。

3）配置交换机 S1 和 S2 各 VLAN 接口的 IP 地址，见表 4-2。

表 4-2

VLAN 10	VLAN 20	VLAN 30	VLAN 100	VLAN 200
192.168.10.1	192.168.20.1	192.168.30.1	192.168.100.1	192.168.100.2

4）PC1-PC3 的网络设置，见表 4-3。

表 4-3

设备	IP 地址	网关	掩码
PC1	192.168.10.10	192.168.10.1	255.255.255.0
PC2	192.168.20.10	192.168.20.1	255.255.255.0
PC3	192.168.30.10	192.168.30.1	255.255.255.0

任务 2：配置交换机 VLAN 信息及虚接口地址。

交换机 S1：

```
S1(config)#vlan 10
S1(config-vlan10)#switchport interface ethernet 1/0/1
Set the port Ethernet1/0/1 access vlan 10 successfully
S1(config-vlan10)#exit
S1(config)#vlan 20
S1(config-vlan20)#switchport interface ethernet 1/0/2
Set the port Ethernet1/0/2 access vlan 20 successfully
S1(config-vlan20)#exit
S1(config)#vlan 100
S1(config-vlan20)#switchport interface ethernet 1/0/24
Set the port Ethernet1/0/24 access vlan 100 successfully
S1(config-vlan20)#exit
S1(config)#interface vlan 10
S1(config-if-vlan10)#ip address 192.168.10.1 255.255.255.0
S1(config-if-vlan10)#exit
```

```
S1(config)#interface vlan 20
S1(config-if-vlan20)#ip address 192.168.20.1 255.255.255.0
S1(config-if-vlan20)#exit
S1(config)#interface vlan 100
S1(config-if-vlan20)#ip address 192.168.100.1 255.255.255.0
S1(config-if-vlan20)#exit
S1(config)#
```

交换机 S2：

```
S2(config)#vlan 30
S2(Config-Vlan30)#switchport interface ethernet 1/1
Set the port Ethernet1/1 access vlan 30 successfully
S2(Config-Vlan30)#exit
S2(config)#vlan 200
S2(Config-Vlan30)#switchport interface ethernet 1/24
Set the port Ethernet1/24 access vlan 200 successfully
S2(Config-Vlan30)#exit
S2(config)# interface vlan 30
S2(Config-if-Vlan30)#ip address 192.168.30.1 255.255.255.0
S2(Config-if-Vlan30)#exit
S2(config)# interface vlan 200
S2(Config-if-Vlan30)#ip address 192.168.100.2 255.255.255.0
S2(Config-if-Vlan30)#exit
S2(config)#
```

任务 3：配置静态路由。

交换机 S1：

```
S1(config)#ip router 0.0.0.0 0.0.0.0 192.168.100.2
S1(config)#show ip router
Codes: K - kernel, C - connected, S - static, R - RIP, B - BGP
       O - OSPF, IA - OSPF inter area
       N1 - OSPF NSSA external type 1, N2 - OSPF NSSA external type 2
       E1 - OSPF external type 1, E2 - OSPF external type 2
       i - IS-IS, L1 - IS-IS level-1, L2 - IS-IS level-2, ia - IS-IS inter area
       * - candidate default

Gateway of last resort is 192.168.100.2 to network 0.0.0.0

S*     0.0.0.0/0 [1/0] via 192.168.100.2, Vlan100   tag:0
C      127.0.0.0/8 is directly connected, Loopback   tag:0
```

```
C           192.168.10.0/24 is directly connected, Vlan10    tag:0
C           192.168.20.0/24 is directly connected, Vlan20    tag:0
C           192.168.100.0/24 is directly connected, Vlan100    tag:0
Total routers are : 4 item(s)
S1(config)#
```

交换机 S2：

```
S2(config)#ip router 0.0.0.0 0.0.0.0 192.168.100.1
S2(config)#show ip router
Codes: K - kernel, C - connected, S - static, R - RIP, B - BGP
        O - OSPF, IA - OSPF inter area
        N1 - OSPF NSSA external type 1, N2 - OSPF NSSA external type 2
        E1 - OSPF external type 1, E2 - OSPF external type 2
        i - IS-IS, L1 - IS-IS level-1, L2 - IS-IS level-2, ia - IS-IS inter area
        * - candidate default

Gateway of last resort is 192.168.100.1 to network 0.0.0.0

S*          0.0.0.0/0 [1/0] via 192.168.100.1, Vlan200
C           127.0.0.0/8 is directly connected, Loopback
C           192.168.30.0/24 is directly connected, Vlan30
C           192.168.100.0/24 is directly connected, Vlan200
Total routers are : 4 item(s)
S2(config)#
```

小张：师父，这里可以使用 RIP 或 OSPF 动态路由实现吗？

李工：当然可以。

任务 4：测试 PC1 与 PC3 的连通性。代码如下。

```
C:\Documents and Settings\Administrator>ping 192.168.30.10

Pinging 192.168.30.10 with 32 bytes of data:

Reply from 192.168.30.10: bytes=32 time=12ms TTL=62
Reply from 192.168.30.10: bytes=32 time<1ms TTL=62
Reply from 192.168.30.10: bytes=32 time<1ms TTL=62
Reply from 192.168.30.10: bytes=32 time<1ms TTL=62

Ping statistics for 192.168.30.10:
    Packets: Sent = 4, Received = 4, Lost = 0 (0% loss),
Approximate round trip times in milli-seconds:
    Minimum = 0ms, Maximum = 0ms, Average = 0ms
```

我们看到，这时 PC1 可以正常与 PC3 通信，PC2 同样可以与 PC3 通信。

任务 5：在交换机 S2 上配置标准 ACL。代码如下。

```
S2(config)#access-list 10 deny 192.168.10.10 0.0.0.255    //拒绝来自 192.168.10.0 网段的数据包
S2(config)#access-list 10 permit any-source               //允许其他任意网段的数据包通过
S2(config)#interface ethernet 1/0/24
S2(config-if-ethernet1/0/24)#ip access-group 10 in        //将 ACL 应用到端口上
S2(config-if-ethernet1/0/24)#exit
S2(config)#
```

李工：小张，这样配置后会出现什么情况呢？

小张：师父，是不是 PC2 可以 Ping 通 PC3，但 PC1 则不能 Ping 通 PC3 吧。

李工：你可以试试。

PC1 Ping PC3：

```
C:\Documents and Settings\Administrator>ping 192.168.30.10

Pinging 192.168.30.10 with 32 bytes of data:

Request timed out.
Request timed out.
Request timed out.
Request timed out.

Ping statistics for 192.168.30.10:
    Packets: Sent = 4, Received = 0, Lost = 4 (100% loss),
```

PC2 Ping PC3：

```
C:\Documents and Settings\Administrator>ping 192.168.30.10

Pinging 192.168.30.10 with 32 bytes of data:

Reply from 192.168.30.10: bytes=32 time=2ms TTL=62
Reply from 192.168.30.10: bytes=32 time<1ms TTL=62
Reply from 192.168.30.10: bytes=32 time<1ms TTL=62
Reply from 192.168.30.10: bytes=32 time<1ms TTL=62

Ping statistics for 192.168.30.10:
    Packets: Sent = 4, Received = 4, Lost = 0 (0% loss),
Approximate round trip times in milli-seconds:
    Minimum = 0ms, Maximum = 1ms, Average = 0ms
```

小张：果然是这样，S2 拒绝了来自 PC1 的数据包。

实训 2　扩展访问控制列表

实训目标

李工：标准 ACL 只能限制源 IP 地址，在目的端拒绝源 IP 地址的访问。而扩展 ACL 的限制权限就很广泛，包括源 IP、目的 IP、服务类型等。下面以限制 Ping 包 ICMP 为例，介绍扩展 ACL 的应用。

实训拓扑

实训拓扑如图 4-2 所示。

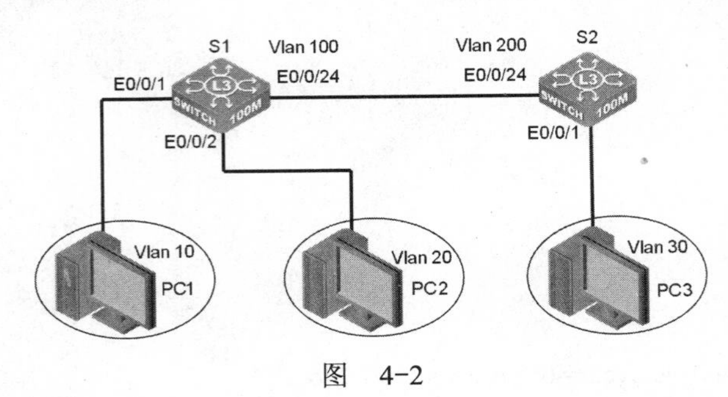

图　4-2

实训任务

任务 1：环境规划。

1）在交换机 A 和交换机 B 上分别划分基于端口的 VLAN，见表 4-4。

表　4-4

交　换　机	VLAN	端 口 成 员
交换机 S1	10	E0/0/1
	20	E0/0/2
	100	E0/0/24
交换机 S2	30	E0/0/1
	200	E0/0/24

2）交换机 S1 和 S2 通过的 24 口级联。

3）配置交换机 S1 和 S2 各 VLAN 接口的 IP 地址，见表 4-5。

表　4-5

VLAN 10	VLAN 20	VLAN 30	VLAN 100	VLAN 200
192.168.10.1	192.168.20.1	192.168.30.1	192.168.100.1	192.168.100.2

4）PC1-PC3 的网络设置，见表 4-6。

表　4-6

设　备	IP 地 址	网　关	掩　码
PC1	192.168.10.10	192.168.10.1	255.255.255.0
PC2	192.168.20.10	192.168.20.1	255.255.255.0
PC3	192.168.30.10	192.168.30.1	255.255.255.0

任务 2：配置交换机 VLAN 信息及虚接口地址。

交换机 S1：

```
S1(config)#vlan 10
S1(config-vlan10)#switchport interface ethernet 1/0/1
Set the port Ethernet1/0/1 access vlan 10 successfully
S1(config-vlan10)#exit
S1(config)#vlan 20
S1(config-vlan20)#switchport interface ethernet 1/0/2
Set the port Ethernet1/0/2 access vlan 20 successfully
S1(config-vlan20)#exit
S1(config)#vlan 100
S1(config-vlan20)#switchport interface ethernet 1/0/24
Set the port Ethernet1/0/24 access vlan 100 successfully
S1(config-vlan20)#exit
S1(config)#interface vlan 10
S1(config-if-vlan10)#ip address 192.168.10.1 255.255.255.0
S1(config-if-vlan10)#exit
S1(config)#interface vlan 20
S1(config-if-vlan20)#ip address 192.168.20.1 255.255.255.0
S1(config-if-vlan20)#exit
S1(config)#interface vlan 100
S1(config-if-vlan20)#ip address 192.168.100.1 255.255.255.0
S1(config-if-vlan20)#exit
S1(config)#
```

交换机 S2：

```
S2(config)#vlan 30
S2(Config-Vlan30)#switchport interface ethernet 1/1
Set the port Ethernet1/1 access vlan 30 successfully
S2(Config-Vlan30)#exit
S2(config)#vlan 200
```

```
S2(Config-Vlan30)#switchport interface ethernet 1/24
Set the port Ethernet1/24 access vlan 200 successfully
S2(Config-Vlan30)#exit
S2(config)# interface vlan 30
S2(Config-if-Vlan30)#ip address 192.168.30.1 255.255.255.0
S2(Config-if-Vlan30)#exit
S2(config)# interface vlan 200
S2(Config-if-Vlan30)#ip address 192.168.100.2 255.255.255.0
S2(Config-if-Vlan30)#exit
S2(config)#
```

任务 3：配置静态路由。

交换机 S1：

```
S1(config)#ip router 0.0.0.0 0.0.0.0 192.168.100.2
S1(config)#show ip router
Codes: K - kernel, C - connected, S - static, R - RIP, B - BGP
       O - OSPF, IA - OSPF inter area
       N1 - OSPF NSSA external type 1, N2 - OSPF NSSA external type 2
       E1 - OSPF external type 1, E2 - OSPF external type 2
       i - IS-IS, L1 - IS-IS level-1, L2 - IS-IS level-2, ia - IS-IS inter area
       * - candidate default

Gateway of last resort is 192.168.100.2 to network 0.0.0.0

S*     0.0.0.0/0 [1/0] via 192.168.100.2, Vlan100   tag:0
C      127.0.0.0/8 is directly connected, Loopback   tag:0
C      192.168.10.0/24 is directly connected, Vlan10   tag:0
C      192.168.20.0/24 is directly connected, Vlan20   tag:0
C      192.168.100.0/24 is directly connected, Vlan100   tag:0
Total router s are : 4 item(s)
S1(config)#
```

交换机 S2：

```
S2(config)#ip router 0.0.0.0 0.0.0.0 192.168.100.1
S2(config)#show ip router
Codes: K - kernel, C - connected, S - static, R - RIP, B - BGP
       O - OSPF, IA - OSPF inter area
       N1 - OSPF NSSA external type 1, N2 - OSPF NSSA external type 2
       E1 - OSPF external type 1, E2 - OSPF external type 2
       i - IS-IS, L1 - IS-IS level-1, L2 - IS-IS level-2, ia - IS-IS inter area
       * - candidate default
```

```
Gateway of last resort is 192.168.100.1 to network 0.0.0.0

S*       0.0.0.0/0 [1/0] via 192.168.100.1, Vlan200
C        127.0.0.0/8 is directly connected, Loopback
C        192.168.30.0/24 is directly connected, Vlan30
C        192.168.100.0/24 is directly connected, Vlan200
Total router s are : 4 item(s)
S2(config)#
```

任务 4：测试 PC1、PC2 到 PC3 的连通性。代码如下。

```
C:\Documents and Settings\Administrator>ping 192.168.30.10

Pinging 192.168.30.10 with 32 bytes of data:

Reply from 192.168.30.10: bytes=32 time=12ms TTL=62
Reply from 192.168.30.10: bytes=32 time<1ms TTL=62
Reply from 192.168.30.10: bytes=32 time<1ms TTL=62
Reply from 192.168.30.10: bytes=32 time<1ms TTL=62

Ping statistics for 192.168.30.10:
    Packets: Sent = 4, Received = 4, Lost = 0 (0% loss),
Approximate round trip times in milli-seconds:
    Minimum = 0ms, Maximum = 0ms, Average = 0ms
```

我们看到，这时 PC1 与 PC2 都可以正常 Ping 通 PC3。

任务 5：在交换机 S1 上配置扩展 ACL。代码如下。

```
S1(config)#access-list 101 deny icmp 192.168.20.10 0.0.0.255 192.168.30.10 0.0.0.255
            //拒绝 192.168.20.0 网段的 ICMP 包访问 192.168.30.0 网段
S1(config)#access-list 101 permit icmp any-source any-destination
            //允许来自其他网段的 ICMP 包访问任意网段
S1(config)#interface ethernet 1/0/1-2
S1(config-if-port-range)#ip access-group 101 in          //将 ACL 应用到端口上
S1(config-if-port-range)#exit
S1(config)#
```

小张：为什么这里要配置在 S1 上呢？

李工：扩展 ACL 应尽量靠近源地址，所以这里把扩展 ACL 配置在 S1 上。而标准 ACL 由于只能限制源 IP 地址，为避免错误的限制，应尽量靠近目的地址，故上例中配置在了 S2 上。

我们现在用 PC1 Ping PC3，应该是通的。代码如下。

```
C:\Documents and Settings\Administrator>ping 192.168.30.10
```

```
Pinging 192.168.30.10 with 32 bytes of data:

Reply from 192.168.30.10: bytes=32 time=1ms TTL=62
Reply from 192.168.30.10: bytes=32 time<1ms TTL=62
Reply from 192.168.30.10: bytes=32 time<1ms TTL=62
Reply from 192.168.30.10: bytes=32 time<1ms TTL=62

Ping statistics for 192.168.30.10:
    Packets: Sent = 4, Received = 4, Lost = 0 (0% loss),
Approximate round trip times in milli-seconds:
    Minimum = 0ms, Maximum = 1ms, Average = 0ms
```

而用 PC2 Ping PC3，则不能 Ping 通。代码如下。

```
C:\Documents and Settings\Administrator>ping 192.168.30.10

Pinging 192.168.30.10 with 32 bytes of data:

Request timed out.
Request timed out.
Request timed out.
Request timed out.

Ping statistics for 192.168.30.10:
    Packets: Sent = 4, Received = 0, Lost = 4 (100% loss),
```

实训 3　三层交换机 DHCP 服务

实训目标

李工：网络中一般都采用 DHCP 作为地址分配的方法，以减轻网络管理员和用户的配置负担。我们可以将支持 DHCP 的交换机配置成 DHCP 服务器，自动为连接它的终端设备分配 IP 地址。

实训拓扑

实训拓扑图如图 4-3 所示。

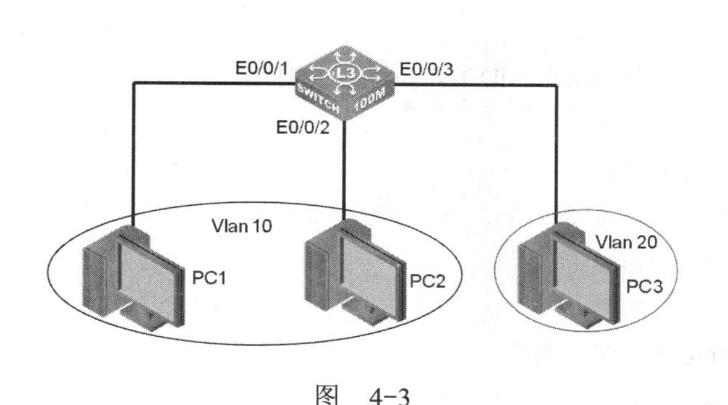

图　4-3

实训任务

任务 1：环境规划。

划分两个 VLAN，见表 4-7。

表　4-7

VLAN	IP 地 址	端 口 成 员
VLAN10	192.168.10.1/24	1-2
VLAN20	192.168.20.1/24	3

配置两个地址池，见表 4-8。

表　4-8

POOLA（192.168.10.0/24）		POOLB（192.168.20.0/24）	
设备 PC1	IP 自动获取	设备 PC3	IP 自动获取
默认网关	192.168.10.1	默认网关	192.168.20.1
DNS 服务器	192.168.1.1	DNS 服务器	192.168.1.1
Lease	8 小时	Lease	1 小时

其中在 VLAN10 中，需将 MAC 地址为 00-0B-4C-9E-2A-1C 的 PC2 分配固定 IP 地址 192.168.10.88。

任务 2：划分 VLAN，加入端口，配置 VLAN 虚接口地址。代码如下。

```
Switch(config)#vlan 10
Switch(config-vlan10)#switchport interface ethernet 1/0/1-2
Set the port Ethernet1/0/1 access vlan 10 successfully
Set the port Ethernet1/0/2 access vlan 10 successfully
Switch(config-vlan10)#exit
Switch(config)#vlan 20
Switch(config-vlan20)#switchport interface ethernet 1/0/3
Set the port Ethernet1/0/3 access vlan 20 successfully
Switch(config-vlan20)#exit
```

```
Switch(config)# interface vlan 10
Switch(config-if-vlan10)#ip address 192.168.10.1 255.255.255.0
Switch(config-if-vlan10)#exit
Switch(config)#interface vlan 20
Switch(config-if-vlan20)#ip address 192.168.20.1 255.255.255.0
Switch(config-if-vlan20)#exit
Switch(config)#
```

任务 3：配置 DHCP。代码如下。

```
Switch(config)#service dhcp
Switch(config)#ip dhcp pool POOLA
Switch(dhcp-poola-config)#network-address 192.168.10.0 24
Switch(dhcp-poola-config)#lease 0 8
Switch(dhcp-poola-config)#default- router r 192.168.10.1
Switch(dhcp-poola-config)#dns-server 192.168.1.1
Switch(dhcp-poola-config)#exit
Switch(config)#ip dhcp pool POOLB
Switch(dhcp-poolb-config)#network-address 192.168.20.0 24
Switch(dhcp-poolb-config)#lease 0 1
Switch(dhcp-poolb-config)#default- router r 192.168.20.1
Switch(dhcp-poolb-config)#dns-server 192.168.1.1
Switch(dhcp-poolb-config)#exit
Switch(config)#
```

任务 4：为特殊的 PC2 配置地址池。代码如下。

```
Switch(config)#ip dhcp excluded-address 192.168.10.80 192.168.10.90
Switch(config)#ip dhcp pool POOLC
Switch(dhcp-poolc-config)#host 192.168.10.88
Switch(dhcp-poolc-config)#hardware-address 00-0B-4C-9E-2A-1C
Switch(dhcp-poolc-config)#default- router r 192.168.10.1
Switch(dhcp-poolc-config)#dns-server 192.168.1.1
Switch(dhcp-poolc-config)#exit
Switch(config)#
```

任务 5：验证，见表 4-9。

表 4-9

设 备	位 置	结 果
PC1	E1/0/1	192.168.10.2/24
PC2	E1/0/2	192.168.10.88/24
PC3	E1/0/3	192.168.20.2/24

实训 4　三层交换机 DHCP 中继

实训目标

李工：当 DHCP 客户机和 DHCP 服务器不在同一个网段时，由 DHCP 中继传递 DHCP 报文。增加 DHCP 中继功能的好处是不必为每个网段都设置 DHCP 服务器，同一个 DHCP 服务器可以为很多个子网的客户机提供网络配置参数，既节约了成本又方便了管理。这就是 DHCP 中继的功能。

实训拓扑

实训拓扑图如图 4-4 所示。

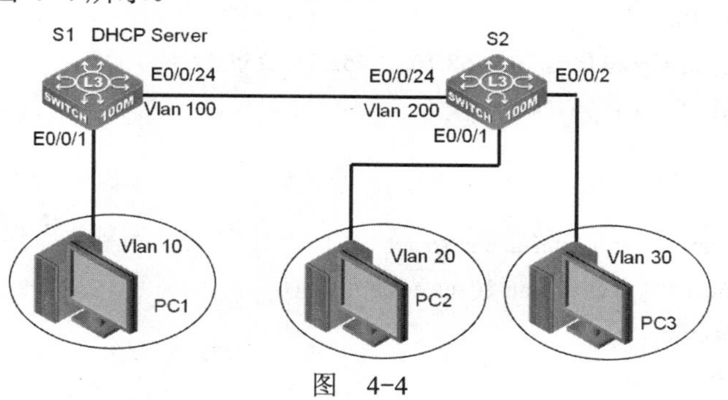

图　4-4

实训任务

任务 1：环境规划。

交换机 S1 为 DHCP 服务器，划分两个 VLAN，见表 4-10。

表　4-10

VLAN	IP 地 址	端 口 成 员
VLAN 10	192.168.10.1/24	1
VLAN 100	192.168.100.1/24	24

为三个 VLAN 分别配置三个地址池。

交换机 S2 开启 DHCP 中继，划分三个 VLAN，见表 4-11。

表　4-11

VLAN	IP 地 址	端 口 成 员
VLAN 20	192.168.20.1/24	1
VLAN 30	192.168.30.1/24	2
VLAN 200	192.168.100.2/24	24

其中，两台交换机通过 24 口相连，并配置为 192.168.100.0 网段。

任务 2：分别在两台交换机上划分 VLAN，加入端口，配置虚接口地址。

交换机 S1：

```
S1(config)#vlan 10
S1(Config-Vlan10)#switchport interface ethernet 1/1
Set the port Ethernet1/1 access vlan 10 successfully
S1(Config-Vlan10)#exit
S1(config)#vlan 100
S1(Config-Vlan100)#switchport interface ethernet 1/24
Set the port Ethernet1/24 access vlan 100 successfully
S1(Config-Vlan100)#exit
S1(config)#int vlan 10
S1(Config-if-Vlan10)#ip address 192.168.10.1 255.255.255.0
S1(Config-if-Vlan10)#exit
S1(config)#int vlan 100
S1(Config-if-Vlan100)#ip address 192.168.100.1 255.255.255.0
S1(Config-if-Vlan100)#exit
```

交换机 S2：

```
S2(config)#vlan 20
S2(config-vlan20)#switchport interface ethernet 1/0/1
Set the port Ethernet1/0/1 access vlan 20 successfully
S2(config-vlan20)#exit
S2(config)#vlan 30
S2(config-vlan30)#switchport interface ethernet 1/0/2
Set the port Ethernet1/0/2 access vlan 30 successfully
S2(config-vlan30)#exit
S2(config)#vlan 200
S2(config-vlan200)#switchport interface ethernet 1/0/24
Set the port Ethernet1/0/24 access vlan 200 successfully
S2(config-vlan200)#exit
S2(config)#int vlan 20
S2(Config-if-Vlan20)#ip address 192.168.20.1 255.255.255.0
S2(Config-if-Vlan20)#exit
S2(config)#int vlan 30
S2(Config-if-Vlan30)#ip address 192.168.30.1 255.255.255.0
S2(Config-if-Vlan30)#exit
S2(config)#int vlan 200
S2(Config-if-Vlan200)#ip address 192.168.100.2 255.255.255.0
S2(Config-if-Vlan200)#exit
```

任务 3：配置静态路由。

交换机 S1：

```
S1(config)#ip router 0.0.0.0 0.0.0.0 192.168.100.2
```

交换机 S2：

```
S2(config)#ip router 0.0.0.0 0.0.0.0 192.168.100.1
```

任务 4：配置交换机 S1 的 DHCP 地址池。代码如下。

```
S1(config)#service dhcp
S1(config)#ip dhcp pool poola
S1(dhcp-poola-config)#network-address 192.168.10.0 24
S1(dhcp-poola-config)#default-router 192.168.10.1
S1(dhcp-poola-config)#dns-server 192.168.1.1
S1(dhcp-poola-config)#exit
S1(config)#ip dhcp pool poolb
S1(dhcp-poolb-config)#network-address 192.168.20.0 24
S1(dhcp-poolb-config)#default-router 192.168.20.1
S1(dhcp-poolb-config)#dns-server 192.168.1.1
S1(dhcp-poolb-config)#exit
S1(config)#ip dhcp pool poolc
S1(dhcp-poolc-config)#network-address 192.168.30.0 24
S1(dhcp-poolc-config)#default-router 192.168.30.1
S1(dhcp-poolc-config)#dns-server 192.168.1.1
S1(dhcp-poolc-config)#exit
```

任务 5：配置交换机 S2 的 DHCP 中继。代码如下。

```
S2(config)#service dhcp
S2(config)#ip forward-protocol udp bootps
S2(config)#int vlan 20
S2(Config-if-vlan20)#ip helper-address 192.168.100.1
S2(Config-if-vlan20)#exit
S2(Config)#int vlan 30
S2(config-if-vlan30)#ip helper-address 192.168.100.1
S2(config-if-vlan30)#exit
S2(config)#
```

任务 6：验证。见表 4-12。

表　4-12

设　　备	位　　置	结　　果
PC1	S1　E1/1	192.168.10.2/24
PC2	S2　E1/0/1	192.168.20.2/24
PC3	S2　E1/0/2	192.168.30.2/24

实训 5　虚拟路由器冗余协议

实训目标

李工：虚拟路由器冗余协议（Virtual Router Redundancy Protocol，VRRP）是一种容错协议，运行于局域网的多台路由器上，它将这几台路由器组织成一台“虚拟”路由器，或称为一个备份组（Standby Group）。在 VRRP 备份组内，总有一台路由器或以太网交换机是活动路由器（Master），它完成“虚拟”路由器的工作；该备份组中其他的路由器或以太网交换机作为备份路由器（Backup，可以不只一台），随时监控 Master 的活动。当原有的 Master 出现故障时，各 Backup 将自动选举出一个新的 Master 来接替其工作，继续为网段内各主机提供路由服务。由于这个选举和接替阶段短暂而平滑，因此，网段内各主机仍然可以正常地使用虚拟路由器，实现不间断地与外界保持通信。

实训拓扑

实训拓扑图如图 4-5 所示。

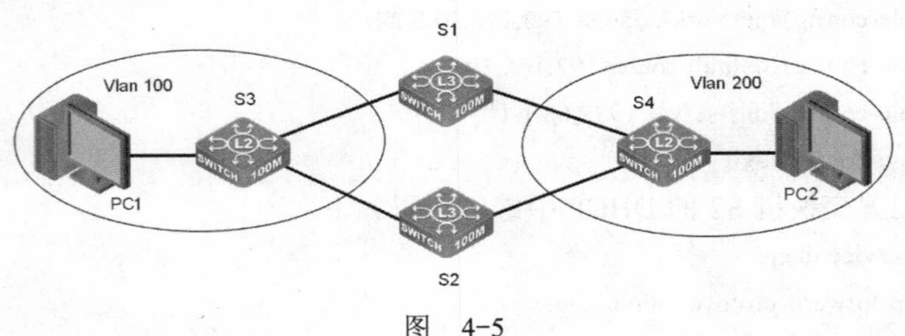

图　4-5

实训任务

任务 1：环境规划。

在 S1 和 S2 上划分 VLAN，见表 4-13。

表　4-13

交 换 机	VLAN	端口成员	IP 地 址
S1	100	1	10.0.0.1/24
	200	24	20.0.0.1/24
S2	100	1	10.0.0.2/24
	200	24	20.0.0.2/24

PC1 和 PC2 的网络设置，见表 4-14。

表 4-14

设 备	IP 地 址	网关（VRRP 虚拟网关）	掩 码
PC1	10.0.0.3	10.0.0.10	255.255.255.0
PC2	20.0.0.3	20.0.0.10	255.255.255.0

任务 2：在交换机 S1 和 S2 上分别划分 VLAN，加入端口，配置虚接口地址。

交换机 S1：

```
S1(config)#vlan 100
S1(config-vlan100)#switchport interface ethernet 1/0/1
Set the port Ethernet1/0/1 access vlan 100 successfully
S1(config-vlan100)#exit
S1(config)#vlan 200
S1(config-vlan200)#switchport interface ethernet 1/0/24
Set the port Ethernet1/0/24 access vlan 200 successfully
S1(config-vlan200)#exit
S1(config)#int vlan 100
S1(config-if-vlan100)#ip address 10.0.0.1 255.255.255.0
S1(config-if-vlan100)#exit
S1(config)#int vlan 200
S1(config-if-vlan200)#ip address 20.0.0.1 255.255.255.0
S1(config-if-vlan200)#exit
S1(config)#
```

交换机 S2：

```
S2(config)#vlan 100
S2(Config-Vlan100)#switchport interface ethernet 1/1
Set the port Ethernet1/1 access vlan 100 successfully
S2(Config-Vlan100)#exit
S2(config)#vlan 200
S2(Config-Vlan200)#switchport interface ethernet 1/24
Set the port Ethernet1/24 access vlan 200 successfully
S2(Config-Vlan200)#exit
S2(config)#int vlan 100
S2(Config-if-Vlan100)#ip address 10.0.0.2 255.255.255.0
S2(Config-if-Vlan100)#exit
S2(config)#int vlan 200
S2(Config-if-Vlan200)#ip address 20.0.0.2 255.255.255.0
S2(Config-if-Vlan200)#exit
S2(config)#
```

任务 3：配置 VRRP。

交换机 S1：

```
S1(config)#router vrrp 1
S1(config-router)#virtual-ip 10.0.0.10
S1(config-router)#priority 150
S1(config-router)#interface vlan 100
S1(config-router)#circuit-failover vlan 200 110
S1(config-router)#enable
S1(config-router)#exit
S1(config)#router vrrp 2
S1(config-router)#virtual-ip 20.0.0.10
S1(config-router)#priority 50
S1(config-router)#interface vlan 200
S1(config-router)#enable
S1(config-router)#exit
S1(config)#
```

验证配置：

```
S1#show vrrp
VrId 1
  State is Master
  Virtual IP is 10.0.0.10 (Not IP owner)
  Interface is Vlan100
  Configured priority is 150, Current priority is 150
  Advertisement interval is 1 sec
  Preempt mode is TRUE
  Circuit failover interface Vlan200, Priority Delta 110, Status UP
VrId 2
  State is Backup
  Virtual IP is 20.0.0.10 (Not IP owner)
  Interface is Vlan200
  Priority is 50
  Advertisement interval is 1 sec
  Preempt mode is TRUE

S1#
```

交换机 S2：

```
S2(config)#router vrrp 1
S2(config-router)#virtual-ip 10.0.0.10
S2(config-router)#priority 50
S2(config-router)#interface vlan 100
S2(config-router)#enable
S2(config-router)#exit
S2(config)#router vrrp 2
S2(config-router)#virtual-ip 20.0.0.10
S2(config-router)#priority 150
```

```
S2(config-router)#interface vlan 200
S2(config-router)#circuit-failover vlan 100 110
S2(config-router)#enable
S2(config-router)#exit
S2(config)#
```

验证配置：

```
S2#show vrrp
VrId 1
 State is Backup
 Virtual IP is 10.0.0.10 (Not IP owner)
 Interface is Vlan100
 Priority is 50
 Advertisement interval is 1 sec
 Preempt mode is TRUE
VrId 2
 State is Master
 Virtual IP is 20.0.0.10 (Not IP owner)
 Interface is Vlan200
 Configured priority is 150, Current priority is 150
 Advertisement interval is 1 sec
 Preempt mode is TRUE
 Circuit failover interface Vlan100, Priority Delta 110, Status UP

S2#
```

任务 4：验证。

PC1 持续 Ping PC2，分别断开 S1 到 S3 及 S1 到 S4 的链路。如图 4-6 所示。

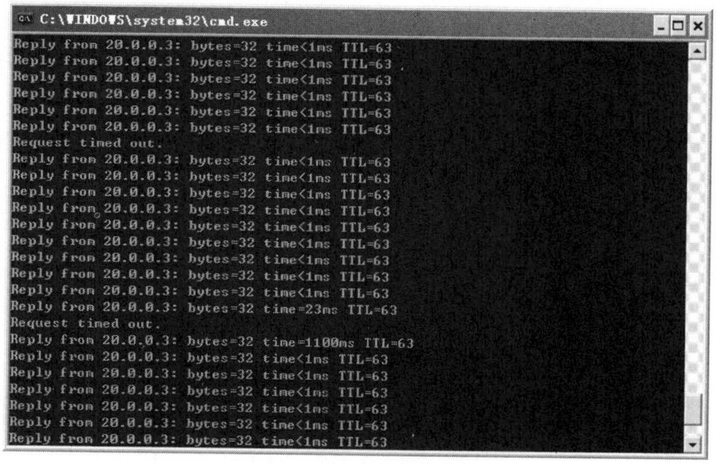

```
C:\WINDOWS\system32\cmd.exe
Reply from 20.0.0.3: bytes=32 time<1ms TTL=63
Reply from 20.0.0.3: bytes=32 time<1ms TTL=63
Reply from 20.0.0.3: bytes=32 time<1ms TTL=63
Reply from 20.0.0.3: bytes=32 time<1ms TTL=63
Reply from 20.0.0.3: bytes=32 time<1ms TTL=63
Reply from 20.0.0.3: bytes=32 time<1ms TTL=63
Request timed out.
Reply from 20.0.0.3: bytes=32 time<1ms TTL=63
Reply from 20.0.0.3: bytes=32 time<1ms TTL=63
Reply from 20.0.0.3: bytes=32 time<1ms TTL=63
Reply from 20.0.0.3: bytes=32 time<1ms TTL=63
Reply from 20.0.0.3: bytes=32 time<1ms TTL=63
Reply from 20.0.0.3: bytes=32 time<1ms TTL=63
Reply from 20.0.0.3: bytes=32 time<1ms TTL=63
Reply from 20.0.0.3: bytes=32 time<1ms TTL=63
Reply from 20.0.0.3: bytes=32 time<1ms TTL=63
Reply from 20.0.0.3: bytes=32 time=23ms TTL=63
Request timed out.
Reply from 20.0.0.3: bytes=32 time=1100ms TTL=63
Reply from 20.0.0.3: bytes=32 time<1ms TTL=63
Reply from 20.0.0.3: bytes=32 time<1ms TTL=63
Reply from 20.0.0.3: bytes=32 time<1ms TTL=63
Reply from 20.0.0.3: bytes=32 time<1ms TTL=63
Reply from 20.0.0.3: bytes=32 time<1ms TTL=63
Reply from 20.0.0.3: bytes=32 time<1ms TTL=63
```

图　4-6

可以看到 VRRP 状态切换时有少量丢包，工作正常。

第2部分

路由器实训

第 5 章　路由器实践基础

实训 1　路由器的基本管理方法

实训目标

小刘中职毕业后经过了集成公司的历练，终于盼来了给运营商做网络的机会，可他师傅告诉他，运营商里最低级别都是企业级的路由器，让他抓紧时间好好学习下路由器的知识，由于下周就要开始项目的实施，小曹被告知务必要在 3 天的时间里搞清楚路由器的带内带外管理方法，并尽可能多地了解路由器的端口特性和配置模式等常识。

实训设备

1）DCR-2655/2659 路由器 1 台。
2）PC 1 台。
3）Console 线揽、交叉双绞线各 1 条。

实训拓扑

实训拓扑图如图 5-1 所示。

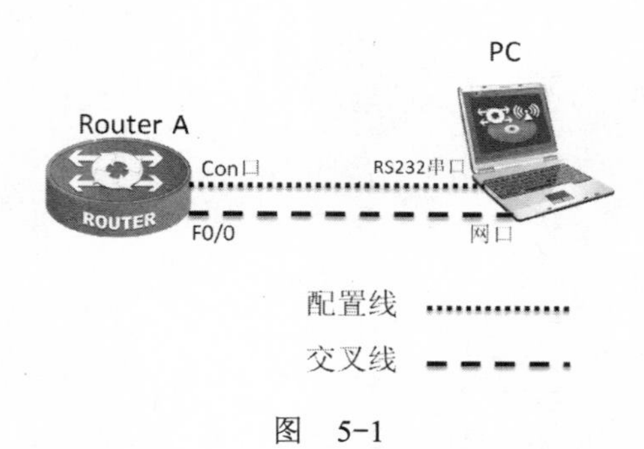

图　5-1

实训任务

任务 1：通过 Console 接口对设备进行初始配置（带外管理方法）。

1）将配置线的一端与路由器的 Console 口相连，另一端与 PC 的串口相连，如图 5-1 所示。

2）在 PC 上运行终端仿真程序。单击“开始”→“程序”→“附件”→“通讯”，运行“超级终端”程序。打开如图 5-2 所示的对话框，设置终端的硬件参数（包括串口号）。

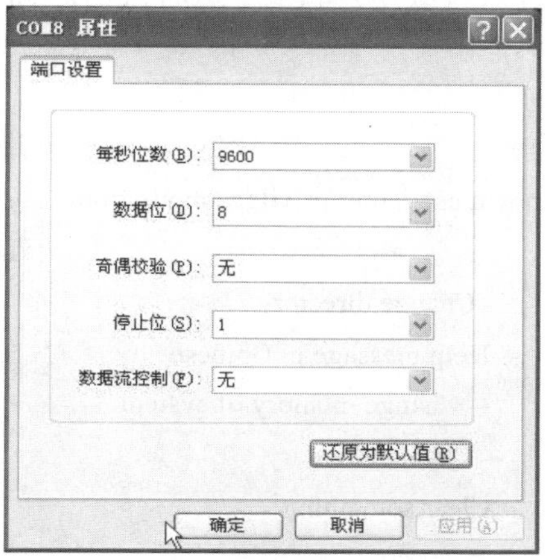

图　5-2

波特率：9600

数据位：8

奇偶校验：无

停止位：1

流控：无

3）路由器加电，超级终端会显示路由器自检信息，自检结束后出现命令提示“Press RETURN to get started”。代码如下。

```
System Bootstrap, Version 0.4.2                    //路由器版本信息
Serial num:8IRTJ610B407000041, ID num:200213
Copyright 2011 by Digital China Networks(BeiJing) Limited
Digital China-DCR-2659 Series DCR-2659                         //路由器型号
The current time: 65024-118-60 2:227:120
Loading DCR26V1.3.3H.bin......                    //加载配置文件
Start Decompress DCR26V1.3.3H.bin
#########################################################################
#########################################################################
#########################################################################
```

```
Decompress 5519234 byte,Please wait system up..
Digital China Networks Limited Internetwork Operating System Software
Digital China-DCR-2659 Series Software , Version 1.3.3H, RELEASE SOFTWARE
System start up OK
Router console 0 is now available
Press RETURN to get started
```

4）按<Enter>键进入用户配置模式。DCR 路由器出厂时没有设定密码，用户按<Enter>键直接进入普通用户模式。可以使用权限允许范围内的命令，需要帮助可以随时输入“？”，输入 enable，按<Enter>键则进入超级用户模式。这时候用户拥有最大权限，可以任意配置，需要帮助可以随时输入“？”。

```
Router>
Router>enable                                              //进入特权模式
Router#Jan  1 00:03:58 Unknown user enter privilege mode from console 0, level = 15
Router#?                                                   //查看可用命令
  cd                  -- Change directory
  chinese             -- Help message in Chinese
  chmem               -- Change memory of system
  chram               -- Change memory
  clear               -- Clear something
  config              -- Enter configurative mode
  connect             -- Open a outgoing connection
  copy                -- Copy configuration or image data
  debug               -- Debugging functions
  delete              -- Delete a file
  dir                 -- List files in flash memory
  disconnect          -- Disconnect an existing outgoing network connection
  download            -- Download with ZMODEM
  enable              -- Turn on privileged commands
  english             -- Help message in English
  enter               -- Turn on privileged commands
  exec-script         -- Execute a script on a port or line
  exit                -- Exit / quit
  format              -- Format file system
  help                -- Description of the interactive help system
  history             -- Look up history
  keepalive           -- Keepalive probe
  look                -- Display memory
 --More—                                          //按<Enter>键可查看更多命令
Router#ch?                                        //使用？帮助
```

```
  chinese                  -- Help message in Chinese
  chmem                    -- Change memory of system
  chram                    -- Change memory
Router#chinese                              //使用中文帮助
Router#?                                    //再次查看可用命令
  cd                       -- 改变当前目录
  chinese                  -- 中文帮助信息
  chmem                    -- 修改系统内存数据
  chram                    -- 修改内存数据
  clear                    -- 清除
  config                   -- 进入配置状态
  connect                  -- 打开一个向外的连接
  copy                     -- 复制配置方案或内存映像
  debug                    -- 分析功能
  delete                   -- 删除一个文件
  dir                      -- 显示闪存中的文件
  disconnect               -- 断开活跃的网络连接
  download                 -- 通过 ZMODEM 协议下载文件
  enable                   -- 进入特权方式
  english                  -- 英文帮助信息
  enter                    -- 进入特权方式
  exec-script              -- 在指定端口运行指定的脚本
  exit                     -- 退回或退出
  format                   -- 格式化文件系统
  help                     -- 交互式帮助系统描述
  history                  -- 查看历史
  keepalive                -- 保活探测
  look                     -- 显示内存数据
--More--
```

任务 2：给相应的接口配置 IP 地址，开启相应的服务，进行带内的管理。

1）设置路由器以太网接口地址并验证连通性。代码如下。

```
Router>enable                              //进入特权模式
Router#Jan  1 00:11:17 Unknown user enter privilege mode from console 0, level = 15
Router#config                              //进入全局配置模式
Router_config#interface f0/0               //进入接口模式
Router_config_f0/0#ip address 192.168.2.1 255.255.255.0    //配置 IP 地址
Router_config_f0/0#no shutdown             //开启端口
Router_config_f0/0#^Z                      // 返回特权模式 （ctrl+Z）
Router#Jan  1 00:12:15 Configured from console 0 by
```

```
Router#show interface f0/0                      //验证
FastEthernet0/0 is up, line protocol is up      //接口和协议都必须 UP
address is 00e0.0f26.0381
  MTU 1500 bytes, BW 100000 kbit, DLY 10 usec
Interface address is 192.168.2.1/24             //接口地址为 192.168.2.1/24
  Encapsulation ARPA
  ARP type: ARPA, ARP timeout 00:03:00
  60 second input rate 151 bits/sec, 0 packets/sec!
  60 second output rate 6 bits/sec, 0 packets/sec!
  Full-duplex, 100Mbit/s, 100BaseTX, 36 ii, 1 oi        //全双工、百兆
    36 packets input, 6394 bytes, 200 rx_freebuf
    Received 0 unicasts, 0 lowmark, 36 ri, 0 input errors
    0 overrun, 0 CRC, 0 framing, 0 busy, 0 long, 0 discard, 0 throttles
    1 packets output, 46 bytes, 210 tx_freebd, 0 output errors
    0 underrun, 0 collisions, 0 late collisions, 0 deferred, 0 reTx expired
    0 resets, 0 lost carrier, 0 no carrier 0 grace stop 0 bus error
    0 output buffer failures, 0 output buffers swapped out 0 tx errors
    0 packets fast-switch
```

2）设置 PC 的 IP 地址并测试连通性，如图 5-3 所示。

3）使用 PING 测试连通性，如图 5-4 所示。

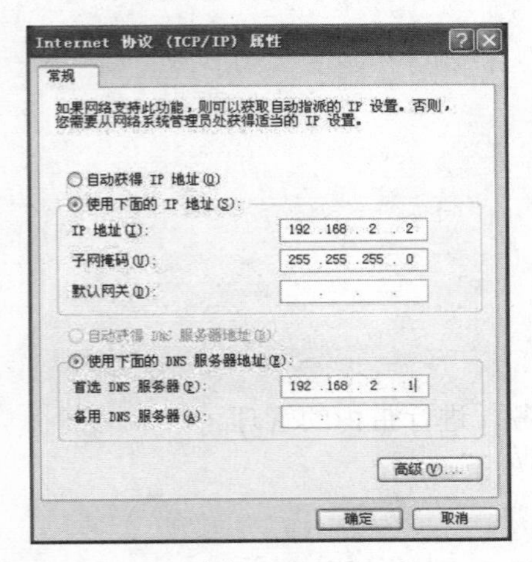

图　5-3

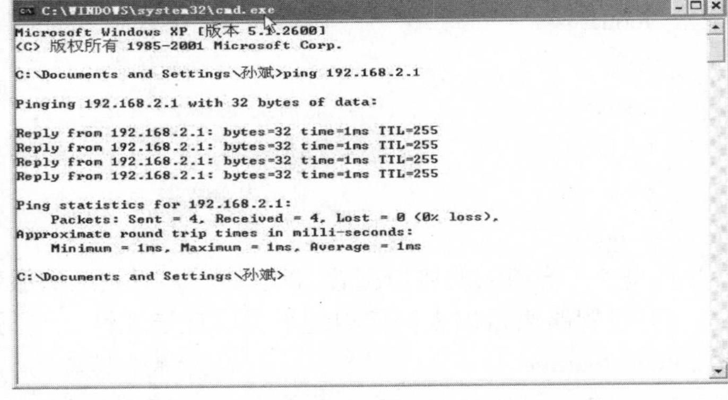

图　5-4

4）在 PC 上 telnet 到路由器。

为了保证安全性，路由器中默认为所有的 telnet 用户必须通过验证才可以进入路由器配置页面，在 DCR 系列路由器中，系统支持使用本地或其他验证数据库对 telnet 用户进行验证，具体过程可参考如下：

设置本地数据库中的用户名，本例使用 dcnu 和密码 dcnu。

```
Router_config #username dcnu password dcnu      //设置本地用户名和密码
Router_config#
```

创建一个新的登录验证方法，名为 login_fortelnet，此方法将使用本地数据库验证。

```
Router_config#aaa authentication login login_fortelnet local      //创建 login_fortelnet 验证，采用 local
Router_config#
```

进入 telnet 进程管理配置模式，配置登录用户使用 login_fortelnet 的验证方法进行验证。

```
Router_config#line vty 0 4
Router_config_line#login authentication login_fortelnet      //在接口下应用
Router_config_line#
```

经过这样的配置，telnet 登录路由器时的过程将如下所示：

```
C:\>telnet 192.168.2.1
User Access Verification
Username: dcnu
Password:
            Welcome to Digital China Multi-Protocol DCR-2659 Series
Router>
```

注意事项和排错

1）在超级终端中的配置是对路由器的操作，这时的 PC 只是输入输出设备。

2）在 telnet 和 web 方式管理时，先测试连通性。

命令解释（见表 5-1）

表　5-1

命　　令	目　　的
interface fastethernet 0/0	进入快速以太网口 F0/0
ip address 192.168.2.2 255.255.255.0	设置 IP 地址 192.168.2.2 子网 255.255.255.0
no shutdown	激活端口
ip http server	启动 HTTP 服务
username dcnu password dcnu	用户名：dcnu 密码：dcnu
aaa authentication login login_forhttp local	创建 login_fortelnet 验证，采用 local

共同思考

1）带内和带外管理方式各有什么优点和缺点？

2）telnet 和 web 的端口号是什么？

实训 2　维护路由器的配置文件

实训目标

小刘通过三天时间的熟悉，终于弄懂了企业级路由器的带内和带外管理方法，可是他又发现了新的问题，就是如何对路由器的软件进行升级，以及如何对路由器的配置文件进行备份和还原，于是他决定利用一天的时间来搞懂这两个问题。

实训设备

1）DCR-2655/2659 路由器 1 台。
2）PC1 台。
3）TFTP、FTP 软件。
4）Console 线揽、交叉双绞线各 1 条。

实训拓扑

实训拓扑图如图 5-5 所示。

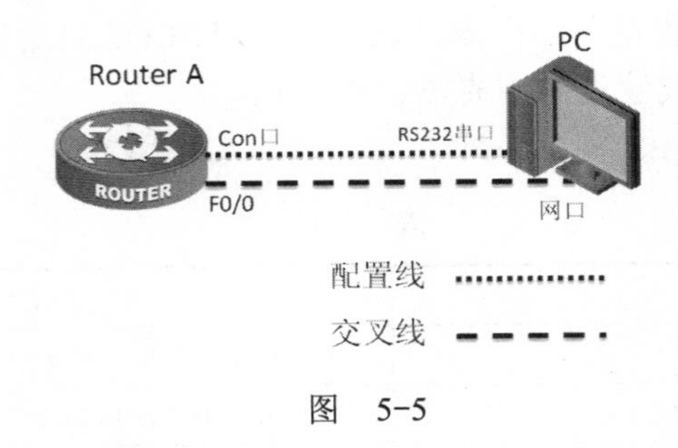

图　5-5

实训任务

任务 1：TFTP 方式（采用 UDP，适合本地操作）。

1）设置 PC 网卡地址为 192.168.2.10，并安装 TFTP SERVER（略，方法参考相关实训）。

2）参照相关实训，设置 DCR-2655/2659 的 F0/0 接口地址为 192.168.2.1，并测试连通性。代码如下。

```
Router>enable                //进入特权模式
Router#config                //进入全局模式
```

```
Router_config#hostname R1        //修改主机名为 R1
R1_config_f0/0#ip address 192.168.2.1 255.255.255.0      //配置接口 IP 地址
R1_config_f0/0#no shutdown        //开启端口
R1_config_f0/0#^Z
R1#Jan   1 00:26:22 Configured from console 0 by

R1#show interface f0/0         //查看端口信息
FastEthernet0/0 is up, line protocol is up
address is 00e0.0f26.0381
  MTU 1500 bytes, BW 100000 kbit, DLY 10 usec
  Interface address is 192.168.2.1/24
  Encapsulation ARPA
  ARP type: ARPA, ARP timeout 00:03:00
  60 second input rate 316 bits/sec, 0 packets/sec!
  60 second output rate 327 bits/sec, 0 packets/sec!
  Full-duplex, 100Mb/s, 100BaseTX, 508 ii, 808 oi
    510 packets input, 38501 bytes, 200 rx_freebuf
    Received 494 unicasts, 0 lowmark, 508 ri, 0 input errors
    0 overrun, 0 CRC, 0 framing, 0 busy, 0 long, 0 discard, 0 throttles
    471 packets output, 35453 bytes, 210 tx_freebd, 0 output errors
    0 underrun, 0 collisions, 0 late collisions, 0 deferred, 0 reTx expired
    0 resets, 0 lost carrier, 0 no carrier 0 grace stop 0 bus error
    0 output buffer failures, 0 output buffers swapped out 0 tx errors
    0 packets fast-switch
R1r#ping 192.168.2.10        //检测与 PC 的连通性
PING 192.168.2.10 (192.168.2.10): 56 data bytes
!!!!!
--- 192.168.2.10 ping statistics ---
5 packets transmitted, 5 packets received, 0% packet loss
round-trip min/avg/max = 2/5/10ms

R1#write              //保存配置文件
Saving current configuration...
OK!
```

3）查看路由器文件，并将配置文件下载到 TFTP 服务器上。代码如下。

```
R1#dir                 //查看路由器文件
Directory of /:
0     DCR26V1.3.3H.bin          <FILE>      5521346      Thu Jan    1 00:00:00 1970
1     Function.map            <FILE>      792462      Thu Jan    1 00:00:00 1970
```

```
2    startup-config    <FILE>    774    Tue Jan  1 00:30:44 2002
//系统默认的配置文件
free space 10420224

Router#copy startup-config tftp          //将配置文件复制到TFTP服务器
Remote-server ip address[]?192.168.2.10      //选择TFTP服务器地址
Destination file name[startup-config]?       //默认文件名（注：这块最好用当天备份时间来对文件重命名，命令为：Destination file name[startup-config]?2012.7.31）
#
TFTP:successfully send 2 blocks ,762 bytes      //备份成功
```

4）打开TFTP目录，使用写字板打开下载后的配置文件，修改机器名，再次使用copy命令上传到路由器中，重新启动后通过show命令观察到机器名已经被修改。代码如下。

```
R1#copy tftp:2012.7.31 startup-config     //从TFTP服务器复制文件2012.7.31，并重命名为startup-config
Remote-server ip address[]?192.168.2.10       //TFTP服务器地址
#
TFTP:successfully receive 2 blocks ,774 bytes      //复制文件成功

R1#reboot            //重新启动路由器
```

任务2：FTP方式（采用TCP，适合远程操作）。

1）设置PC网卡地址为192.168.2.10，并安装TFTP SERVER（略，方法参考相关实训）。

2）参照相关实训，设置DCR-2655/2659的F0/0接口地址为192.168.2.1，并测试连通性。代码如下。

```
Router>enable                  //进入特权模式
Router#config                  //进入全局模式
Router_config#hostname R1         //修改主机名为R1
R1_config_f0/0#ip address 192.168.2.1 255.255.255.0     //配置接口IP地址
R1_config_f0/0#no shutdown        //开启端口
R1_config_f0/0#^Z
R1#Jan  1 00:26:22 Configured from console 0 by

R1#show interface f0/0         //查看端口信息
FastEthernet0/0 is up, line protocol is up
address is 00e0.0f26.0381
  MTU 1500 bytes, BW 100000 kbit, DLY 10 usec
  Interface address is 192.168.2.1/24
  Encapsulation ARPA
  ARP type: ARPA, ARP timeout 00:03:00
  60 second input rate 316 bits/sec, 0 packets/sec!
```

```
    60 second output rate 327 bits/sec, 0 packets/sec!
    Full-duplex, 100Mb/s, 100BaseTX, 508 ii, 808 oi
      510 packets input, 38501 bytes, 200 rx_freebuf
      Received 494 unicasts, 0 lowmark, 508 ri, 0 input errors
      0 overrun, 0 CRC, 0 framing, 0 busy, 0 long, 0 discard, 0 throttles
      471 packets output, 35453 bytes, 210 tx_freebd, 0 output errors
      0 underrun, 0 collisions, 0 late collisions, 0 deferred, 0 reTx expired
      0 resets, 0 lost carrier, 0 no carrier 0 grace stop 0 bus error
      0 output buffer failures, 0 output buffers swapped out 0 tx errors
      0 packets fast-switch
R1r#ping 192.168.2.10          //检测与 PC 的连通性
PING 192.168.2.10 (192.168.2.10): 56 data bytes
!!!!!
--- 192.168.2.10 ping statistics ---
5 packets transmitted, 5 packets received, 0% packet loss
round-trip min/avg/max = 2/5/10ms

R1#write                 //保存配置文件
Saving current configuration...
OK!
```

3）查看路由器文件，并将配置文件下载到 TFTP 服务器上。代码如下。

```
R1#dir                     //查看路由器文件
Directory of /:
0     DCR26V1.3.3H.bin        <FILE>      5521346     Thu Jan   1 00:00:00 1970
1     Function.map            <FILE>      792462      Thu Jan   1 00:00:00 1970
2     startup-config          <FILE>      774         Tue Jan   1 00:30:44 2002
//系统默认的启动文件
free space 10420224

Router#copy startup-config ftp            //将配置文件复制到 FTP 服务器
ftp user name[anonymous]?dcnu           //输入 FTP 用户名
ftp user password[anonymous]?123456            //输入 FTP 用户密码
Remote-server ip address[]?192.168.2.10         //FTP 服务器地址
Destination file name[startup-config]?          //默认文件名
#
FTP:successfully send 2 blocks ,762 bytes         //备份成功
```

4）打开 FTP 目录，使用写字板打开下载后的配置文件，修改机器名，再次使用 copy 命令上传到路由器中，重新启动后通过 show 命令观察到机器名已经被修改。代码如下。

```
R1#copy ftp startup-config 192.168.2.10         //从 FTP 服务器复制文件
```

```
ftp user name[anonymous]?dcnu
ftp user password[anonymous]?123456
Source file name[]?
#
FTP:successfully receive 2 blocks ,774 bytes        //复制文件成功

R1#reboot              //重新启动路由器
```

任务 3：启动到 monitor 模式。

当路由器的软件被破坏而无法启动时，可以在启动过程中按<Ctrl+Break>键，启动到 monitor 模式中，使用 ZMODEM 方式恢复文件。所谓 ZMODEM 方式是从路由器的 Console 端口以波特率规定的速率通过 PC 的串口传输文件的一种方式，不需要网线，但速度很慢，本实训以 TFTP 作为传输协议。

1）将路由器重启，在启动过程中按<Ctrl+Break>键，启动到 MONITOR。代码如下。

```
System Bootstrap, Version 0.4.2
Serial num:8IRTJ610B407000041, ID num:200213
Copyright 2011 by Digital China Networks(BeiJing) Limited
Digital China-DCR-2659 Series DCR-2659
The current time: 65024-118-60 2:227:120


        Welcome to Digital China Multi-Protocol DCR-2659 Series Router
monitor#                             //进入 MONITOR 模式
```

2）设置 IP 地址，测试连通性。代码如下。

```
monitor#ip address 192.168.2.1 255.255.255.0          //配置 IP 地址
monitor#ping 192.168.2.10                              //测试连通性
Ping 192.168.2.10 with 48 bytes of data:
Reply from 192.168.2.10: bytes=48 time=10ms TTL=128
Reply from 192.168.2.10: bytes=48 time=10ms TTL=128
Reply from 192.168.2.10: bytes=48 time=10ms TTL=128
Reply from 192.168.2.10: bytes=48 time=10ms TTL=128
4 packets sent, 4 packets received
round trip min/avg/max = 10/10/10 ms

monitor#
```

3）PC 启动 TFTP 服务，开始 copy 过程。代码如下。

```
monitor#copy tftp flash: 192.168.2.10          //从 TFTP 服务器复制文件
Source file name[]?startup-config              //需要复制的文件名
Destination file name[startup-config]?         //默认文件名
#
TFTP:successfully receive 2 blocks ,778 bytes
```

```
monitor#
```

4）重新启动设备。代码如下。

```
monitor#reboot
Do you want to reboot the router(y/n)?y
Please wait..


System Bootstrap, Version 0.4.2
Serial num:8IRTJ610B407000041, ID num:200213
Copyright 2011 by Digital China Networks(BeiJing) Limited
Digital China-DCR-2659 Series DCR-2659
The current time: 65024-118-60 2:227:120
Loading DCR26V1.3.3H.bin......
Start Decompress DCR26V1.3.3H.bin
################################################################################
################################################################################
################################################################################
Decompress 5519234 byte,Please wait system up..
Digital China Networks Limited Internetwork Operating System Software
Digital China-DCR-2659 Series Software , Version 1.3.3H, RELEASE SOFTWARE
System start up OK


Router console 0 is now available


Press RETURN to get started
                Welcome to DCR Multi-Protocol 1700 Series Router
Router>
```

任务 4：恢复遗失的密码。

当密码遗忘，可以进入 monitor 模式，执行以下操作清除密码。

```
monitor#nopasswd
```

任务 5：使用 TFTP 维护路由器的原系统备份并升级成新版本。

1）使用 show version 命令查看当前系统版本。代码如下。

```
Router#show version                    //查看路由器系统信息
Digital China Networks Limited Internetwork Operating System Software
DCR-2659 Series Software, Version 1.3.3H (MIDDLE), RELEASE SOFTWARE
Copyright 2011 by Digital China Networks(BeiJing) Limited
Compiled: 2011-01-19 15:23:40 by system, Image text-base: 0x6004
ROM: System Bootstrap, Version 0.4.2                    //系统引导版本信息
Serial num:8IRTJ610B407000041, ID num:200213
System image file is "DCR26V1.3.3H.bin"                 //配置文件
```

```
Digital China-DCR-2659 (PowerPC) Processor            //路由器型号
65536K bytes of memory,16384K bytes of flash          //内存容量
Router uptime is 0:00:00:57, The current time: 2002-01-01 00:00:57
Slot 0: SCC Slot
  Port 0: 10/100Mbps full-duplex Ethernet             //百兆全双工以太网口
  Port 1: 2M full-duplex Serial                       //2 兆全双工串口
  Port 2: 2M full-duplex Serial
  Port 3: 1000Mbit/s full-duplex Ethernet             //千兆全双工以太网口
  Port 4: 1000Mbit/s full-duplex Ethernet
  Port 5: 1000Mbit/s full-duplex Ethernet
  Port 6: 1000Mbit/s full-duplex Ethernet
Router#
```

2）使用 dir 命令察看当前系统文件名称。代码如下。

```
Router#dir                   //使用 dir 命令查看系统文件
Directory of /:
0    DCR26V1.3.3H.bin    <FILE>    5521346    Thu Jan  1 00:00:00 1970
1    Function.map        <FILE>    792462     Thu Jan  1 00:00:00 1970
2    startup-config      <FILE>    778        Thu Jan  1 00:00:00 1970
free space 10420224
Router#
```

3）确保 TFTP 服务器与路由器连通。代码如下。

```
Router#ping 192.168.2.10
PING 192.168.2.10 (192.168.2.10): 56 data bytes
!!!!!
--- 192.168.2.10 ping statistics ---
5 packets transmitted, 5 packets received, 0% packet loss
round-trip min/avg/max = 0/0/0 ms
Router#
```

4）开启 TFTP 服务器，使用 copy file tftp 命令备份系统文件。代码如下。

```
Router#copy flash:DCR26V1.3.3H.bin tftp         //将配置文件备份到 TFTP 服务器
Remote-server ip address[]?192.168.2.10         //TFTP 服务器地址
Destination file name[DCR26V1.3.3H.bin]?        //默认文件名
##################################################################
##################################################################
##################################################################
##################################################################
##################################################################
##################################################################
##################################################################
```

```
#######################################################################
#########################################################
TFTP:successfully send 10784 blocks ,5521346 bytes
Router#
```

5）在 TFTP 服务器中察看保存的文件。

打开 TFTP 服务器的主目录，将看到文件，如图 5-6 所示。

图　5-6

6）在神州数码网站下载路由器新版本到 TFTP 服务器主目录中。

下载地址是：http://www.dcnetworks.com.cn/cn/download/，找到合适的设备型号开始下载，将下载文件保存到 TFTP 服务器的主目录中。

7）使用 copy tftp file 命令升级新版本。代码如下。

```
Router#copy tftp flash:                   //从 TFTP 服务器复制文件
Source file name[]?DCR26V1.3.3H.bin
                                          //目标文件名（此处文件名只为举例，请根据实际情况输入）
Remote-server ip address[]?192.168.2.10        //TFTP 服务器地址
Destination file name[DCR26V1.3.3H.bin]?       //默认文件名
#######################################################################
#######################################################################
#######################################################################
#######################################################################
#######################################################################
#######################################################################
#######################################################################
#######################################################################
#######################################################################
#########################################################
TFTP:successfully receive 10784 blocks ,5521346 bytes
Router#
```

注意事项和排错

1）路由器以太网口和 PC 直接相连的时候，使用交叉双绞线。

2）关闭 PC 上的防火墙。

3）在实际工作中，通常使用日期或功能等标明配置文件。

4）使用 ZMODEM 方式，当文件比较大的时候，比较耗时。

共同思考

1）请问 TFTP 和 FTP 这两种方式有什么区别？
2）在 monitor 模式下，为什么不能使用 TFTP 方式？

实训 3　路由器直连路由的配置

实训目标

小刘通过之前的学习，已经掌握了企业级路由器的基本管理方法，现在他准备进入实际配置阶段，于是，他开始学习直连路由的配置。

实训设备

1）DCR2655/2659 路由器 1 台。
2）交叉双绞线 2 条，配置线 1 条。
3）PC2 台。

实训拓扑

实训拓扑图如图 5-7 所示。

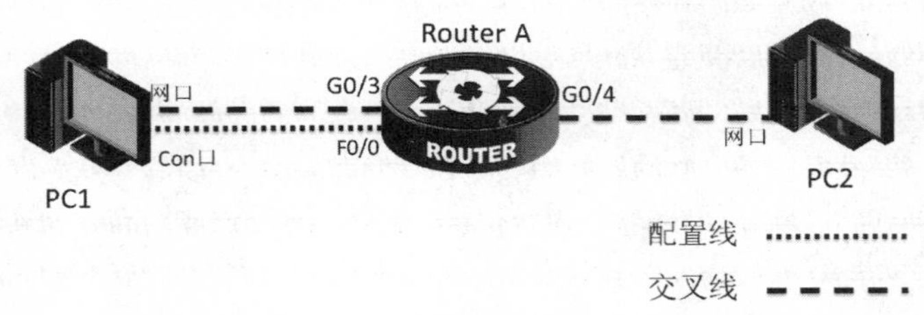

图　5-7

实训任务

任务 1：小刘先分别将 PC1 和 PC2 的以太网口接到了路由器的 G0/3 和 G0/4 口，然后配置 PC 的 IP 地址。见表 5-2。

表　5-2

主　机　名	IP　地　址	子　网　掩　码
PC1	192.168.3.10	255.255.255.0
PC2	192.168.4.10	255.255.255.0

任务 2：PC 互 PING，检测连通性，如图 5-8、图 5-9 所示。

```
选定 C:\WINDOWS\system32\cmd.exe
Microsoft Windows XP [版本 5.1.2600]
(C) 版权所有 1985-2001 Microsoft Corp.

C:\Documents and Settings\Administrator>ping 192.168.4.10

Pinging 192.168.4.10 with 32 bytes of data:

Destination host unreachable.
Destination host unreachable.
Destination host unreachable.
Destination host unreachable.

Ping statistics for 192.168.4.10:
    Packets: Sent = 4, Received = 0, Lost = 4 (100% loss),

C:\Documents and Settings\Administrator>
```

图　5-8

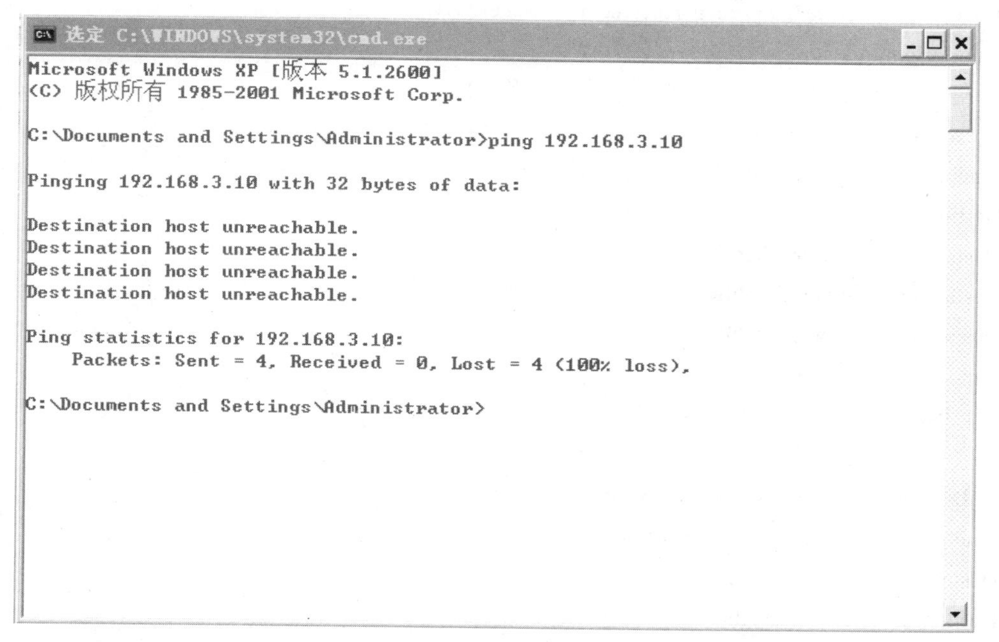

图　5-9

此时，小刘发现两台 PC 并不能互通，经过分析，他发现路由器的接口地址还没有配置，于是小刘开始着手配置路由器接口地址。

任务 3：配置路由器接口地址，并检测互通性。代码如下。

```
Router#config
Router_config#interface g0/3
Router_config_g0/3#ip address 192.168.3.1 255.255.255.0
Router_config_g0/3#exit
Router_config#interface G0/4
Router_config_g0/4#ip address 192.168.4.1 255.255.255.0
Router_config_g0/4#exit
Router_config#exit
Router#ping 192.168.3.10
PING 192.168.3.10 (192.168.3.10): 56 data bytes
!!!!!
--- 192.168.3.10 ping statistics ---
5 packets transmitted, 5 packets received, 0% packet loss
round-trip min/avg/max = 0/0/0 ms
Router#ping 192.168.4.10
PING 192.168.4.10 (192.168.4.10): 56 data bytes
!!!!!
--- 192.168.4.10 ping statistics ---
5 packets transmitted, 5 packets received, 0% packet loss
round-trip min/avg/max = 0/0/0 ms
```

任务 4：PC 互 PING，检测连通性，代码如下。

```
C:\Documents and Settings\Administrator>ping 192.168.4.10
Pinging 192.168.4.10 with 32 bytes of data:
Destination host unreachable.
Destination host unreachable.
Destination host unreachable.
Destination host unreachable.
Ping statistics for 192.168.4.10:
    Packets: Sent = 4, Received = 0, Lost = 4 (100% loss),
```

```
C:\Documents and Settings\Administrator>ping 192.168.3.10
Pinging 192.168.4.10 with 32 bytes of data:
Destination host unreachable.
Destination host unreachable.
Destination host unreachable.
Destination host unreachable.
Ping statistics for 192.168.3.10:
    Packets: Sent = 4, Received = 0, Lost = 4 (100% loss),
```

此时，小刘发现两台 PC 还是不能互相通信，经过仔细检查，他发现虽然已经配置了 PC 以及路由器接口的 IP 地址，但没有给 PC 配置网关，于是他按照下面的表格又重新为 PC 配置了 IP，并重新进行了检测连通性的检测。见表 5-3。

表 5-3

主 机 名	IP 地 址	子网掩码	网 关
PC1	192.168.3.10	255.255.255.0	192.168.3.1
PC2	192.168.4.10	255.255.255.0	192.168.4.1

检测连通性。代码如下。

```
C:\Documents and Settings\Administrator>ping 192.168.4.10
Pinging 192.168.3.10 with 32 bytes of data:
Reply from 192.168.3.10: bytes=32 time<1ms TTL=64
Reply from 192.168.3.10: bytes=32 time<1ms TTL=64
Reply from 192.168.3.10: bytes=32 time<1ms TTL=64
Reply from 192.168.3.10: bytes=32 time<1ms TTL=64
Ping statistics for 192.168.3.10:
    Packets: Sent = 4, Received = 4, Lost = 0 (0% loss),
Approximate round trip times in milli-seconds:
    Minimum = 0ms, Maximum = 0ms, Average = 0ms
```

```
C:\Documents and Settings\Administrator>ping 192.168.3.10
Pinging 192.168.3.10 with 32 bytes of data:
Reply from 192.168.3.10: bytes=32 time<1ms TTL=64
Reply from 192.168.3.10: bytes=32 time<1ms TTL=64
Reply from 192.168.3.10: bytes=32 time<1ms TTL=64
Reply from 192.168.3.10: bytes=32 time<1ms TTL=64
Ping statistics for 192.168.3.10:
    Packets: Sent = 4, Received = 4, Lost = 0 (0% loss),
Approximate round trip times in milli-seconds:
    Minimum = 0ms, Maximum = 0ms, Average = 0ms
```

注意事项和排错

1）本实训中并未对中心路由器配置任何路由，而它却连通了两侧的网络段，并且只需要在两侧设备中配置到达非直连网络的默认路由，即可完成连通性。

2）本实训的结论是对路由器而言，是否能够将数据转发到正确的地点，完全取决于其路由表的覆盖面，只有路由表中存在的表项才可以将数据按照路由表的指示传递出去，而对于路由表中没有标明出口的网络，路由器则处于无知状态，对于目的地址是未知网络的数据，路由器通通丢弃。

共同思考

该实验中 PC1 到 PC2 的报文转发流程是怎样的？

实训 4　路由器单臂路由的配置

实训目标

路由器的以太网端口通常用来连接企业的局域网络，在很多时候内网又划分了多个 VLAN，这些 VLAN 的用户都需要从一个出口访问外网。这个统一的出口在交换机中以 IEEE 802.1Q 的方式封装，就意味着数据从这个出口访问对端设备时，必须能够识别并区分对待来自多个 VLAN 的数据才可以保证链路的正常通信。路由器以太网口在此时应该如何配置呢？这就是本实训需要解决的问题。

实训设备

1）DCS 二层交换机 1 台。
2）DCR2655/2659 路由器 1 台。
3）直通双绞线 3 根，配置线 2 根。
4）PC2 台。

实训拓扑

实训拓扑图如图 5-10 所示。

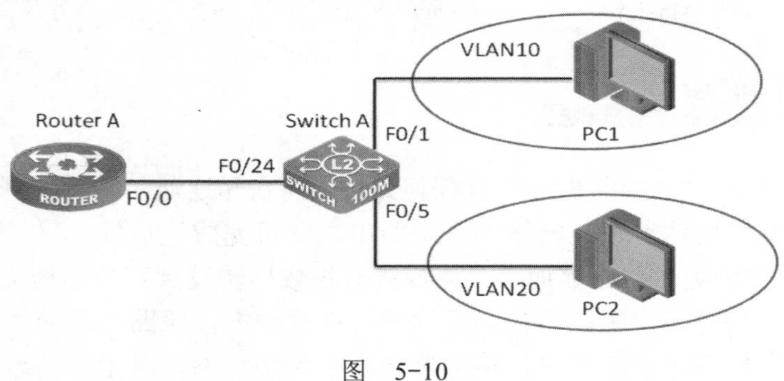

图　5-10

实训任务

任务 1：配置交换机的 VLAN 及其成员端口，设置 24 端口的 trunk 属性，配置 PVID 等。代码如下。

```
DCS-3950-26C#
DCS-3950-26C#config
DCS-3950-26C(config)#vlan 10          //划分 VLAN 10
DCS-3950-26C(config-vlan10)#switchport interface e0/0/1-4
                        //将 E0/0/1、E0/0/2、E0/0/3、E0/0/4 四个端口加入到 VLAN10 中
Set the port Ethernet0/0/1 access vlan 10 successfully
Set the port Ethernet0/0/2 access vlan 10 successfully
Set the port Ethernet0/0/3 access vlan 10 successfully
Set the port Ethernet0/0/4 access vlan 10 successfully
DCS-3950-26C(config-vlan10)#exit
DCS-3950-26C(config)#vlan 20     //划分 VLAN20
DCS-3950-26C(config-vlan20)#switchport interface e0/0/5-8
                        //将 E0/0/5、E0/0/6、E0/0/7、E0/0/8 四个端口加入到 VLAN10 中
Set the port Ethernet0/0/5 access vlan 20 successfully
Set the port Ethernet0/0/6 access vlan 20 successfully
Set the port Ethernet0/0/7 access vlan 20 successfully
Set the port Ethernet0/0/8 access vlan 20 successfully
DCS-3950-26C(config-vlan20)#exit
DCS-3950-26C(config)#interface e0/0/24    //进入 E0/0/24 端口
DCS-3950-26C(config-if-ethernet0/0/24)#switchport mode trunk //将端口设为 TRUNK 模式
Set the port Ethernet0/0/24 mode Trunk successfully
DCS-3950-26C(config-if-ethernet0/0/24)#switchport trunk allowed vlan all    //允许所有 VLAN 通过
set the Trunk port Ethernet0/0/24 allowed vlan successfully
DCS-3950-26C(config-if-ethernet0/0/24)#
```

任务 2：为路由器创建以太网接口的子接口，并在子接口上配置 VID 和对应的 IP 地址等。代码如下。

```
Router_config#interface f0/0.1     //进入路由器子接口
Router_config_f0/0.1#ip address 192.168.1.1 255.255.255.0    //为子接口添加 IP
Router_config_f0/0.1#encapsulation dot1Q 10    //为 VLAN10 的数据包封装 802.1Q 协议
Router_config_f0/0.1#exit
Router_config#interface f0/0.2     //进入路由器子接口
Router_config_f0/0.2#ip address 192.168.2.1 255.255.255.0     //为子接口添加 IP
Router_config_f0/0.2#encapsulation dot1Q 20   //为 VLAN20 的数据包封装 802.1Q 协议
Router_config_f0/0.2#exit
Router_config#
```

任务 3：连接 PC 配置默认网关为路由器对应 VLAN 的接口地址，测试连通性。分别为 PC1 和 PC2 配置 IP 地址，并互 PING。见表 5-4。

表 5-4

主 机 名	IP 地 址	子 网 掩 码	默 认 网 关
PC1	192.168.1.10	255.255.255.0	192.168.1.1
PC2	192.168.2.10	255.255.255.0	192.168.2.1

代码如下。

```
C:\Documents and Settings\Administrator>ping 192.168.2.10
Pinging 192.168.2.10 with 32 bytes of data:
Reply from 192.168.2.10: bytes=32 time<1ms TTL=63
Reply from 192.168.2.10: bytes=32 time<1ms TTL=63
Reply from 192.168.2.10: bytes=32 time<1ms TTL=63
Reply from 192.168.2.10: bytes=32 time<1ms TTL=63
Ping statistics for 192.168.2.10:
    Packets: Sent = 4, Received = 4, Lost = 0 (0% loss),
Approximate round trip times in milli-seconds:
    Minimum = 0ms, Maximum = 0ms, Average = 0ms

C:\Documents and Settings\Administrator>ping 192.168.1.10
Pinging 192.168.1.10 with 32 bytes of data:
Reply from 192.168.1.10: bytes=32 time<1ms TTL=63
Reply from 192.168.1.10: bytes=32 time<1ms TTL=63
Reply from 192.168.1.10: bytes=32 time<1ms TTL=63
Reply from 192.168.1.10: bytes=32 time<1ms TTL=63
Ping statistics for 192.168.1.10:
    Packets: Sent = 4, Received = 4, Lost = 0 (0% loss),
Approximate round trip times in milli-seconds:
    Minimum = 0ms, Maximum = 0ms, Average = 0ms
```

主要命令详解（见表 5-5）

表 5-5

命 令	目 的
switchport interface ethernet 0/0/1-4	把端口 0/0/1 至 0/0/4 都加入指定 vlan
switchport trunk vlan-allowed all	指定的 trunk 口，所有 vlan 可以通过

第6章 路由器路由技术基础

实训1 路由器静态路由的配置

实训目标

小杜是一家集成公司的网络管理员，最近公司在本市又租下了一个办公地点，公司行政部门的同事已经装修得差不多了，目前新办公地点和旧地点之间的网络还相互独立，他们希望能尽快让两部分的网络连通，就好启动新的办公地点了。小杜盘点了下现有设备，发现可用的就是两台旧路由器，他又问了下那边的上网情况，发现新大楼和旧楼都是同一个开发商的产业，因此互联网也都是通过开发商提供的宽带入口接入的。小杜决定把旧路由器用上，两边使用静态路由完成互联互通，经过师傅同意后，小杜开始了工作。

实训设备

1）DCR2655/2659 路由器 3 台。
2）CR-V35FC/CR-V35MT 各 1 条。
3）交叉双绞线 3 条。
4）PC 1 台。

实训拓扑

实训拓扑图如图 6-1 所示。

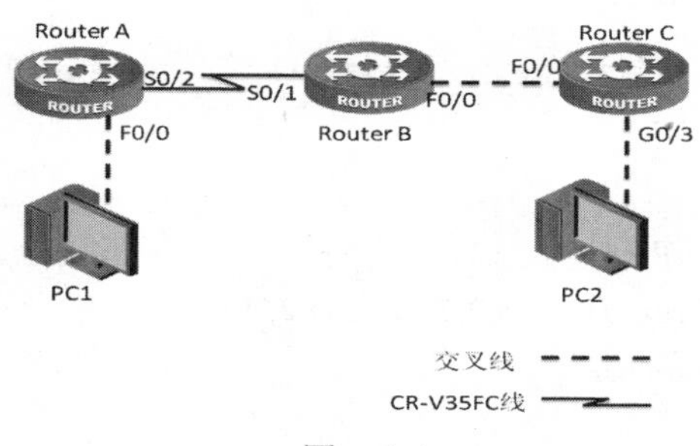

图 6-1

实训任务

任务 1：参照前面的实训方法，按照表 6-1 配置所有接口的 IP 地址，保证所有接口全部是 up 状态，测试连通性。

表 6-1

Router A		Router B		Router C	
S0/2(DCE)	192.168.1.1	S0/1(DTE)	192.168.1.2	F0/0	192.168.2.2
F0/0	192.168.0.1	F0/0	192.168.2.1	G0/3	192.168.3.1

Router A：

```
Router A_config#
Router A_config#interface s0/2                    //进入 S0/2 端口
Router A_config_s0/2#ip address 192.168.1.1 255.255.255.0     //配置接口 IP 地址
Router A_config_s0/2#no shutdown                  //开启端口
Router A_config_s0/2#physical-layer speed 64000    //配置时钟，V35 线的 DCE 端必须要配置时钟
Router A_config_s0/2#Jan   1 00:01:08 Line on Interface Serial0/2, changed to up
Jan   1 00:01:09 Line protocol on Interface Serial0/2, change state to up
Router A_config_s0/2#exit
Router A_config#interface f0/0                    //进入 f0/0 端口
Router A_config_f0/0#ip address 192.168.0.1 255.255.255.0    //配置对口地址
Router A_config_f0/0#no shutdown       //开启端口
Router A_config_f0/0#exit
```

Router B：

```
Router B_config#
Router B _config#interface s0/1
Router B _config_s0/1#ip address 192.168.1.2 255.255.255.0
Router B _config_s0/1#no shutdown
Router B _config_s0/1#exit
Router B _config#interface f0/0
Router B _config_f0/0#ip address 192.168.2.1 255.255.255.0
Router B _config_f0/0#no shutdown
Router B _config_f0/0#exit
```

Router C：

```
Router C_config#
Router C_config#interface f0/0
Router C_config_f0/0#ip address 192.168.2.2 255.255.255.0
Router C_config_f0/0#no shutdown
Router C_config_f0/0#exit
Router C_config#interface g0/3
```

```
Router C_config_g0/3#ip address 192.168.3.1 255.255.255.0
Router C_config_g0/3#no shutdown
Router C_config_g0/3#exit
Router C_config#
```

任务 2：查看路由器 A 的路由表。代码如下。

```
Router A#show ip router
Codes: C - connected, S - static, R - RIP, B - BGP, BC - BGP connected
       D - BEIGRP, DEX - external BEIGRP, O - OSPF, OIA - OSPF inter area
       ON1 - OSPF NSSA external type 1, ON2 - OSPF NSSA external type 2
       OE1 - OSPF external type 1, OE2 - OSPF external type 2
       DHCP - DHCP type, L1 - IS-IS level-1, L2 - IS-IS level-2
VRF ID: 0
C        192.168.0.0/24      is directly connected, FastEthernet0/0
C        192.168.1.0/24      is directly connected, Serial0/2
```

任务 3：查看路由器 B 的路由表。代码如下。

```
Router B#show ip router
Codes: C - connected, S - static, R - RIP, B - BGP, BC - BGP connected
       D - BEIGRP, DEX - external BEIGRP, O - OSPF, OIA - OSPF inter area
       ON1 - OSPF NSSA external type 1, ON2 - OSPF NSSA external type 2
       OE1 - OSPF external type 1, OE2 - OSPF external type 2
       DHCP - DHCP type, L1 - IS-IS level-1, L2 - IS-IS level-2
VRF ID: 0
C        192.168.1.0/24        is directly connected, Serial0/1
C        192.168.2.0/24        is directly connected, FastEthernet0/0
```

任务 4：查看路由器 C 的路由表。代码如下。

```
Router C#show ip router
Codes: C - connected, S - static, R - RIP, B - BGP, BC - BGP connected
       D - BEIGRP, DEX - external BEIGRP, O - OSPF, OIA - OSPF inter area
       ON1 - OSPF NSSA external type 1, ON2 - OSPF NSSA external type 2
       OE1 - OSPF external type 1, OE2 - OSPF external type 2
       DHCP - DHCP type, L1 - IS-IS level-1, L2 - IS-IS level-2
VRF ID: 0
C        192.168.2.0/24          is directly connected, FastEthernet0/0
C        192.168.3.0/24          is directly connected, GigaEthernet0/3
```

任务 5：在路由器 A 上 PING 路由器 C。代码如下。

```
Router A#ping 192.168.2.2
PING 192.168.2.2 (192.168.2.2): 56 data bytes
.....
--- 192.168.2.2 ping statistics ---
```

```
5 packets transmitted, 0 packets received, 100% packet loss          //不通
```

任务 6：在路由器 A 上配置静态路由。代码如下。

```
Router A#config
Router A_config#ip router 192.168.2.0 255.255.255.0 192.168.1.2//配置目标网段和下一跳
Router A_config#ip router 192.168.3.0 255.255.255.0 192.168.1.2
```

任务 7：查看路由表。代码如下。

```
Router A#show ip router
Codes: C - connected, S - static, R - RIP, B - BGP, BC - BGP connected
       D - BEIGRP, DEX - external BEIGRP, O - OSPF, OIA - OSPF inter area
       ON1 - OSPF NSSA external type 1, ON2 - OSPF NSSA external type 2
       OE1 - OSPF external type 1, OE2 - OSPF external type 2
       DHCP - DHCP type, L1 - IS-IS level-1, L2 - IS-IS level-2
VRF ID: 0
C      192.168.0.0/24      is directly connected, FastEthernet0/0
C      192.168.1.0/24      is directly connected, Serial0/2
S      192.168.2.0/24      [1,0] via 192.168.1.2(on Serial0/2)
S      192.168.3.0/24      [1,0] via 192.168.1.2(on Serial0/2)
```

任务 8：配置路由器 B 的静态路由并查看路由表。代码如下。

```
Router B#config
Router B_config#ip router 192.168.0.0 255.255.255.0 192.168.1.1
Router B_config#ip router 192.168.3.0 255.255.255.0 192.168.1.1
Router B_config#exit
Router B#Jan   1 00:22:04 Configured from console 0 by UNKNOWN
Router B#show ip router
Codes: C - connected, S - static, R - RIP, B - BGP, BC - BGP connected
       D - BEIGRP, DEX - external BEIGRP, O - OSPF, OIA - OSPF inter area
       ON1 - OSPF NSSA external type 1, ON2 - OSPF NSSA external type 2
       OE1 - OSPF external type 1, OE2 - OSPF external type 2
       DHCP - DHCP type, L1 - IS-IS level-1, L2 - IS-IS level-2
VRF ID: 0
S      192.168.0.0/24      [1,0] via 192.168.1.1(on Serial0/1)
C      192.168.1.0/24      is directly connected, Serial0/1
C      192.168.2.0/24      is directly connected, FastEthernet0/0
S      192.168.3.0/24      [1,0] via 192.168.1.1(on Serial0/1)
```

任务 9：配置路由器 C 的静态路由并查看路由表。代码如下。

```
Router C#config
Router C_config#ip router 192.168.0.0 255.255.0.0 192.168.2.1          //采用超网的方法
Router C_config#^Z
```

```
Router C#show ip router
Codes: C - connected, S - static, R - RIP, B - BGP
        D - DEIGRP, DEX - external DEIGRP, O - OSPF, OIA - OSPF inter area
        ON1 - OSPF NSSA external type 1, ON2 - OSPF NSSA external type 2
        OE1 - OSPF external type 1, OE2 - OSPF external type 2
S       192.168.0.0/16     [1,0] via 192.168.2.1(on FastEthernet0/0)     //注意子网掩码为 16 位
C       192.168.2.0/24     is directly connected, FastEthernet0/0
C       192.168.3.0/24     is directly connected, GigaEthernet0/3
```

任务 10：测试。代码如下。

```
Router-C#ping 192.168.0.1
PING 192.168.0.1 (192.168.0.1): 56 data bytes
!!!!!                                                          //成功
--- 192.168.0.1 ping statistics ---
5 packets transmitted, 5 packets received, 0% packet loss
round-trip min/avg/max = 30/32/40 ms
```

注意事项和排错

1）非直连的网段都要配置路由。
2）以太网接口要接主机或交换机才能 up。
3）串口注意 DCE 和 DTE 的问题。

命令详解（见表 6-2）

表　6-2

命　　令	目　　的
ip router 192.168.2.0 255.255.255.0 192.168.1.2	静态路由，把 192.168.2.0/24 网段流入的数据，从下一跳为 192.168.1.2 的端口流出

共同思考

1）什么情况下可以采用路由器 C 的超网配置方法？
2）静态路由有什么优势？什么情况下使用？
3）路由器 B 如果不配置任何静态路由，会影响哪些网段间的互通？

实训 2　静态路由掩码最长匹配

实训目标

面对当前 IP 地址资源匮乏的形式，很多企业采用子网的方式设置局域网环境，如果在

三层设备中为每一个子网提供一条确切的路由，当然是最标准的路由设置方法，但这样也会造成三层设备的路由表变大，因此查询效率受到一定的影响。比较合理的做法就是将可能合并成一个大网的子网在路由表中进行汇总，使它们通过一条路由查询到出口。这样做，查询效率会提升，但同时也要求必须对全网的子网分布做合理规划，否则会因为路由表的查询原则造成网络的连通问题。本实训集中讨论关于路由表掩码最长匹配原则，主要针对VLSM环境进行讨论。

实训设备

1）DCR2655/2659路由器2台。
2）交叉双绞线3条。
3）PC2台。

实训拓扑

实训拓扑图如图6-2所示。

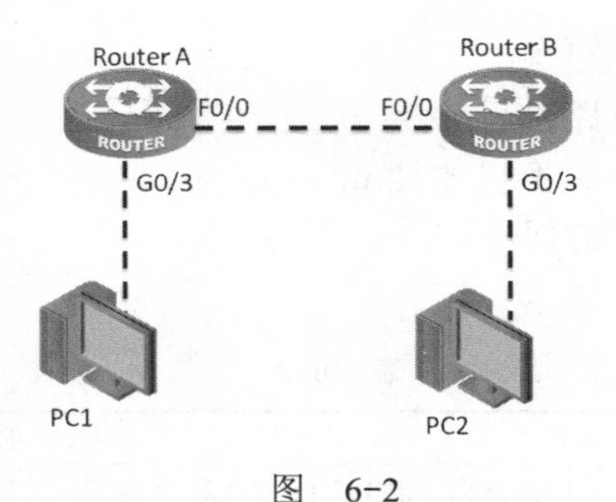

图 6-2

实训任务

任务1：参照前面实训和图6-2所示的拓扑图，按表6-3配置接口。

表 6-3

主 机 名	F0/0	G0/3	PC 地 址	默 认 网 关
Router A	10.1.1.1	10.1.3.1		
Router B	10.1.1.2	10.1.2.1		
PC1			10.1.3.10/24	10.1.3.1
PC2			10.1.2.10/24	10.1.2.1

Router A：

```
Router#
Router#config                    //全局配置模式
```

```
Router_config#hostname RouterA              //更改主机名
Router A _config#interface f 0/0            //进入 f0/0 端口
Router A _config_f0/0#ip address 10.1.1.1 255.255.255.0    //配置接口地址
Router A _config_f0/0#no shutdown           //开启端口
Router A _config_f0/0#exit
Router A _config#interface g 0/3
Router A _config_g0/3#ip add 10.1.3.1 255.255.255.0
Router A _config_g0/3#no shutdown
Router A _config_ g0/3#exit
Router A _config#write                      //保存配置
Saving current configuration...
OK!
Router A _config#
```

Router B：

```
Router_config#hostname Router B
Router B _config#interface f 0/0
Router B _config_f0/0#ip add 10.1.1.2 255.255.255.0
Router B _config_f0/0#exit
Router B _config#ping 10.1.1.1              //检测与 Router A 的连通性
PING 10.1.1.1 (10.1.1.1): 56 data bytes
!!!!!
--- 10.1.1.1 ping statistics ---
5 packets transmitted, 5 packets received, 0% packet loss
round-trip min/avg/max = 0/0/0 ms
Router B _config#interface g 0/3
Router B _config_g0/3#ip add 10.1.2.1255.255.255.0
Router B _config_g0/3#no shutdown
Router B _config_g0/3#exit
Router B _config#wr
Saving current configuration...
OK!
Router B _config#
```

注意：以上已经验证了两台路由器之间单条链路的连通性。

任务 2：配置静态路由。

Router A：

```
Router A_config#ip router 10.0.0.0255.0.0.010.1.1.2          //配置静态路由
Router A_config#ip router 10.1.2.0255.255.255.010.1.3.1     //配置静态路由
```

（注：此处指明的下一跳路由是错误的，目的是通过最长匹配原则让路由器将数据发往错误的地点从而引起不通。）

Router B：

```
Router B _config#ip router 10.0.0.0255.0.0.010.1.1.1          //配置静态路由
Router B _config#ip router 10.1.3.0255.255.255.010.1.2.1      //配置静态路由
```

（注：此处指明的下一跳路由是错误的，目的是通过最长匹配原则让路由器将数据发往错误的地点从而引起不通。）

任务 3：查看路由表。

Router A：

```
Router A#show ip router
Codes: C - connected, S - static, R - RIP, B - BGP, BC - BGP connected
        D - BEIGRP, DEX - external BEIGRP, O - OSPF, OIA - OSPF inter area
        ON1 - OSPF NSSA external type 1, ON2 - OSPF NSSA external type 2
        OE1 - OSPF external type 1, OE2 - OSPF external type 2
        DHCP - DHCP type, L1 - IS-IS level-1, L2 - IS-IS level-2
VRF ID: 0
S       10.0.0.0/8          [1,0] via 10.1.1.2(on FastEthernet0/0)
C       10.1.1.0/24         is directly connected, FastEthernet0/0
S       10.1.2.0/24         [1,0] via 10.1.3.1 (on GigaEthernet0/3)
C       10.1.3.0/24         is directly connected, GigaEthernet0/3
```

Router B：

```
Router B#show ip router
Codes: C - connected, S - static, R - RIP, B - BGP, BC - BGP connected
        D - BEIGRP, DEX - external BEIGRP, O - OSPF, OIA - OSPF inter area
        ON1 - OSPF NSSA external type 1, ON2 - OSPF NSSA external type 2
        OE1 - OSPF external type 1, OE2 - OSPF external type 2
        DHCP - DHCP type, L1 - IS-IS level-1, L2 - IS-IS level-2
VRF ID: 0
S       10.0.0.0/8          [1,0] via 10.1.1.1(on FastEthernet0/0)
C       10.1.1.0/24         is directly connected, FastEthernet0/0
C       10.1.2.0/24         is directly connected, GigaEthernet0/3
S       10.1.3.0/24         [1,0] via 10.1.2.10(on GigaEthernet0/3)
```

任务 4：测试连通性。

PC1：

```
C:\>ipconfig
Windows IP Configuration
Ethernet adapter 本地连接:
        Connection-specific DNS Suffix   . :
        IP Address. . . . . . . . . . . . : 10.1.3.10
        Subnet Mask . . . . . . . . . . . : 255.255.255.0
        Default Gateway . . . . . . . . . : 10.1.3.1
```

```
C:\>ping 10.1.3.1
Pinging 10.1.3.1 with 32 bytes of data:
Reply from 10.1.2.1: bytes=32 time<1ms TTL=255
Reply from 10.1.2.1: bytes=32 time<1ms TTL=255
Reply from 10.1.2.1: bytes=32 time<1ms TTL=255
Reply from 10.1.2.1: bytes=32 time<1ms TTL=255
Ping statistics for 10.1.3.1:
    Packets: Sent = 4, Received = 4, Lost = 0 (0% loss),
Approximate round trip times in milli-seconds:
Minimum = 0ms, Maximum = 0ms, Average = 0ms
```

```
C:\>ping 10.1.2.1
Pinging 10.1.2.1 with 32 bytes of data:
Request timed out.
Request timed out.
Request timed out.
Request timed out.
Ping statistics for 10.1.2.1:
Packets: Sent = 4, Received = 0, Lost = 4 (100% loss),
```

此时看到 PC1 到 10.1.2.1 是连不通的，进一步跟踪路径得到：

```
C:\>tracert 10.1.2.10
Tracing router to 10.1.2.10 over a maximum of 30 hops
  1    <1 ms    <1 ms    <1 ms   10.1.3.1
  2    *        *        *       Request timed out.
  3  ^C
C:\>
```

此时查看到问题出在 10.1.2.1 的下一步无法到达。

注意：PC2 情况类似，此处略。

任务 5：问题分析和解决。

从第 4 步了解到是因为 10.1.3.1 的路由出了问题所以不能连通，因此回顾 Router A 的路由表如下：

```
S    10.1.2.0/24    [1,0] via 10.1.3.1 (on GigaEthernet 0/3)
S    10.0.0.0/8     [1,0] via 10.1.1.1(on FastEthernet0/0)
```

由于目标为 10.1.2.0 网络的成员，因此根据最长匹配的原则，这里从 G0/3 口发送出去，显然它找不到正确的位置了，解决上述问题，有两种办法：

① 将目标地址改为其他网段，比如 10.1.4.0，这样通过路由表，RA 会将数据从 f0/0 发送出去，因此可以到达，但这需要更改 RB 的 G0/3 接口地址为 10.1.4.1，PC2 地址也需

做相应修改，此步骤做课后练习。

② 更改静态路由配置，具体修改方法如下：

----- Router A ------删除 10.1.2.0 的路由

----- Router B ------删除 10.1.3.0 的路由

在设备中实现如下：

Router A：

```
Router A_config#no ip router 10.1.2.0 255.255.255.0 10.1.3.1
Router A _config#sh ip router
Codes: C - connected, S - static, R - RIP, B - BGP, BC - BGP connected
        D - DEIGRP, DEX - external DEIGRP, O - OSPF, OIA - OSPF inter area
        ON1 - OSPF NSSA external type 1, ON2 - OSPF NSSA external type 2
        OE1 - OSPF external type 1, OE2 - OSPF external type 2
        DHCP - DHCP type
VRF ID: 0
S       10.0.0.0/8          [1,0] via 10.1.1.1(on FastEthernet0/0)
C       10.1.1.0/24         is directly connected, FastEthernet0/0
C       10.1.2.0/24         is directly connected, GigaEthernet0/3
```

Router B：

```
Router B_config #no ip router 10.1.3.0 255.255.255.0 10.1.2.1
Router B_config #show ip router
Codes: C - connected, S - static, R - RIP, B - BGP, BC - BGP connected
        D - DEIGRP, DEX - external DEIGRP, O - OSPF, OIA - OSPF inter area
        ON1 - OSPF NSSA external type 1, ON2 - OSPF NSSA external type 2
        OE1 - OSPF external type 1, OE2 - OSPF external type 2
        DHCP - DHCP type
VRF ID: 0
S       10.0.0.0/8          [1,0] via 10.1.1.1(on FastEthernet0/0)
C       10.1.1.0/24         is directly connected, FastEthernet0/0
C       10.1.2.0/24         is directly connected, GigaEthernet0/3
```

再次测试连通性：

```
C:\>ping 10.1.3.10
Pinging 10.1.3.10 with 32 bytes of data:
Reply from 10.1.3.10: bytes=32 time=2ms TTL=126
Reply from 10.1.3.10: bytes=32 time<1ms TTL=126
Reply from 10.1.3.10: bytes=32 time<1ms TTL=126
Reply from 10.1.3.10: bytes=32 time<1ms TTL=126
Ping statistics for 10.1.3.10:
    Packets: Sent = 4, Received = 4, Lost = 0 (0% loss),
```

```
Approximate round trip times in milli-seconds:
Minimum = 0ms, Maximum = 2ms, Average = 0ms
```

此时已经可以连通。

注意事项和排错

本实训第1步将设备hostname改为与拓扑图一致的状态，有助于实训的顺利进行。

本实训最终设置的静态路由的掩码并非最优方案，只是本实训的测试目的，请正常设置静态路由时不要完全效仿。

如果实训过程中发生静态路由无法写入路由表的情况，则查看静态路由指向的下一跳所对应端口是否处于UP状态。

实训3　路由器RIP的配置

实训目标

小宋是一所学校的网络管理员，最近学校要进行一次网络改造，在规划网络时，小宋发现改造后的网络中路由器较多，手工配置静态路由会带来很大的工作负担，同时，在不太稳定的网络环境里，手工修改表不现实，于是他考虑采用动态路由协议。本实训就帮助我们来理解如何配置动态路由，同时理解路由协议的工作过程。

实训设备

1）DCR2655/2659路由器3台。

2）CR-V35FC1/CR-V35MT1各1条。

3）交叉双绞线3条。

4）PC 2台。

实训拓扑

实训拓扑图如图6-3所示。

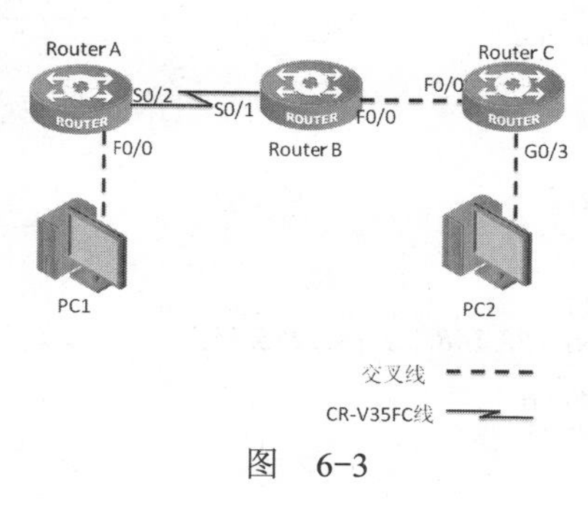

图　6-3

实训任务（见表6-4）

任务1：参照前面实训，按照表6-4配置所有接口的IP地址，保证所有接口全部是UP状态，测试连通性。

表 6-4

Router A		Router B		Router C	
S0/2(DCE)	192.168.1.1	S0/1 (DTE)	192.168.1.2	F0/0	192.168.2.2
F0/0	192.168.0.1	F0/0	192.168.2.1	G0/3	192.168.3.1

Router A：

```
Router A_config#
Router A_config#interface s0/2                    //进入 S0/2 端口
Router A_config_s0/2#ip address 192.168.1.1 255.255.255.0     //配置接口 IP 地址
Router A_config_s0/2#no shutdown                  //开启端口
Router A_config_s0/2#physical-layer speed 64000     //配置时钟，V35 线的 DCE 端必须要配置时钟
Router A_config_s0/2#Jan  1 00:01:08 Line on Interface Serial0/2, changed to up
Jan  1 00:01:09 Line protocol on Interface Serial0/2, change state to up
Router A_config_s0/2#exit
Router A_config#interface f0/0                    //进入 f0/0 端口
Router A_config_f0/0#ip address 192.168.0.1 255.255.255.0    //配置端口地址
Router A_config_f0/0#no shutdown        //开启端口
Router A_config_f0/0#exit
```

Router B：

```
Router B_config#
Router B _config#interface s0/1
Router B _config_s0/1#ip address 192.168.1.2 255.255.255.0
Router B _config_s0/1#no shutdown
Router B _config_s0/1#exit
Router B _config#interface f0/0
Router B _config_f0/0#ip address 192.168.2.1 255.255.255.0
Router B _config_f0/0#no shutdown
Router B _config_f0/0#exit
```

Router C：

```
Router C_config#
Router C_config#interface f0/0
Router C_config_f0/0#ip address 192.168.2.2 255.255.255.0
Router C_config_f0/0#no shutdown
Router C_config_f0/0#exit
```

```
Router C_config#interface g0/3
Router C_config_g0/3#ip address 192.168.3.1 255.255.255.0
Router C_config_g0/3#no shutdown
Router C_config_g0/3#exit
Router C_config#
```

任务 2：查看路由器 A 的路由表。

```
Router A#show ip router
Codes: C - connected, S - static, R - RIP, B - BGP, BC - BGP connected
       D - BEIGRP, DEX - external BEIGRP, O - OSPF, OIA - OSPF inter area
       ON1 - OSPF NSSA external type 1, ON2 - OSPF NSSA external type 2
       OE1 - OSPF external type 1, OE2 - OSPF external type 2
       DHCP - DHCP type, L1 - IS-IS level-1, L2 - IS-IS level-2
VRF ID: 0
C       192.168.0.0/24     is directly connected, FastEthernet0/0
C       192.168.1.0/24     is directly connected, Serial0/2
```

任务 3：查看路由器 B 的路由表。

```
Router B#show ip router
Codes: C - connected, S - static, R - RIP, B - BGP, BC - BGP connected
       D - BEIGRP, DEX - external BEIGRP, O - OSPF, OIA - OSPF inter area
       ON1 - OSPF NSSA external type 1, ON2 - OSPF NSSA external type 2
       OE1 - OSPF external type 1, OE2 - OSPF external type 2
       DHCP - DHCP type, L1 - IS-IS level-1, L2 - IS-IS level-2
VRF ID: 0
C       192.168.1.0/24       is directly connected, Serial0/1
C       192.168.2.0/24       is directly connected, FastEthernet0/0
```

任务 4：查看路由器 C 的路由表。

```
Router C#show ip Router
Codes: C - connected, S - static, R - RIP, B - BGP, BC - BGP connected
       D - BEIGRP, DEX - external BEIGRP, O - OSPF, OIA - OSPF inter area
       ON1 - OSPF NSSA external type 1, ON2 - OSPF NSSA external type 2
       OE1 - OSPF external type 1, OE2 - OSPF external type 2
       DHCP - DHCP type, L1 - IS-IS level-1, L2 - IS-IS level-2
VRF ID: 0
C       192.168.2.0/24          is directly connected, FastEthernet0/0
C       192.168.3.0/24          is directly connected, GigaEthernet0/3
```

任务 5：在路由器 A 上 PING 路由器 C。

```
Router A#ping 192.168.2.2
PING 192.168.2.2 (192.168.2.2): 56 data bytes
```

```
.....
--- 192.168.2.2 ping statistics ---
5 packets transmitted, 0 packets received, 100% packet loss          //不通
```

任务 6：在路由器 A 上配置 RIP 并查看路由表。

```
Router A_config#
Router A_config#router rip                              //启动 RIP 协议
Router A_config_rip#network 192.168.0.0                 //宣告网段
Router A_config_rip#network 192.168.1.0
Router A_config_rip#^Z
Router A#Jan   1 00:11:37 Configured from console 0 by
Router A#show ip router
Codes: C - connected, S - static, R - RIP, B - BGP, BC - BGP connected
       D - BEIGRP, DEX - external BEIGRP, O - OSPF, OIA - OSPF inter area
       ON1 - OSPF NSSA external type 1, ON2 - OSPF NSSA external type 2
       OE1 - OSPF external type 1, OE2 - OSPF external type 2
       DHCP - DHCP type, L1 - IS-IS level-1, L2 - IS-IS level-2
VRF ID: 0
C      192.168.0.0/24      is directly connected, FastEthernet0/0
C      192.168.1.0/24      is directly connected, Serial0/2
```

注意到并没有出现 RIP 学习到的路由。

任务 7：在路由器 B 上配置 RIP 并查看路由表。

```
Router B_config#router rip
Router B_config_rip#network 192.168.1.0
Router B_config_rip#network 192.168.2.0
Router B_config_rip#^Z
Router B#Jan   1 00:15:31 Configured from console 0 by
Router B#
Router B#show ip router
Codes: C - connected, S - static, R - RIP, B - BGP, BC - BGP connected
       D - BEIGRP, DEX - external BEIGRP, O - OSPF, OIA - OSPF inter area
       ON1 - OSPF NSSA external type 1, ON2 - OSPF NSSA external type 2
       OE1 - OSPF external type 1, OE2 - OSPF external type 2
       DHCP - DHCP type, L1 - IS-IS level-1, L2 - IS-IS level-2
VRF ID: 0
R      192.168.0.0/24      [120,1] via 192.168.1.1(on Serial0/1)   //从路由器 A 学习到路由
C      192.168.1.0/24      is directly connected, Serial0/1
C      192.168.2.0/24      is directly connected, FastEthernet0/0
Router B#
```

任务 8：在路由器 C 上配置 RIP 并查看路由表。

```
Router-C_config#router rip
Router-C_config_rip#network 192.168.2.0
Router-C_config_rip#network 192.168.3.0
Router-C_config_rip#^Z
Router C#show ip router
Codes: C - connected, S - static, R - RIP, B - BGP, BC - BGP connected
       D - BEIGRP, DEX - external BEIGRP, O - OSPF, OIA - OSPF inter area
       ON1 - OSPF NSSA external type 1, ON2 - OSPF NSSA external type 2
       OE1 - OSPF external type 1, OE2 - OSPF external type 2
       DHCP - DHCP type, L1 - IS-IS level-1, L2 - IS-IS level-2
VRF ID: 0
R      192.168.1.0/24      [120,1] via 192.168.2.1(on FastEthernet0/0)
C      192.168.2.0/24      is directly connected, FastEthernet0/0
C      192.168.3.0/24      is directly connected, GigaEthernet0/3
```

任务 9：再次查看 A 和 B 的路由表。

```
Router A#show ip router
Codes: C - connected, S - static, R - RIP, B - BGP, BC - BGP connected
       D - BEIGRP, DEX - external BEIGRP, O - OSPF, OIA - OSPF inter area
       ON1 - OSPF NSSA external type 1, ON2 - OSPF NSSA external type 2
       OE1 - OSPF external type 1, OE2 - OSPF external type 2
       DHCP - DHCP type, L1 - IS-IS level-1, L2 - IS-IS level-2
VRF ID: 0
C      192.168.1.0/24      is directly connected, Serial0/2
R      192.168.2.0/24      [120,1] via 192.168.1.2(on Serial0/2)
R      192.168.3.0/24      [120,2] via 192.168.1.2(on Serial0/2)
```

```
Router B#show ip router
Codes: C - connected, S - static, R - RIP, B - BGP, BC - BGP connected
       D - BEIGRP, DEX - external BEIGRP, O - OSPF, OIA - OSPF inter area
       ON1 - OSPF NSSA external type 1, ON2 - OSPF NSSA external type 2
       OE1 - OSPF external type 1, OE2 - OSPF external type 2
       DHCP - DHCP type, L1 - IS-IS level-1, L2 - IS-IS level-2
VRF ID: 0
C      192.168.1.0/24      is directly connected, Serial0/1
C      192.168.2.0/24      is directly connected, FastEthernet0/0
R      192.168.3.0/24      [120,1] via 192.168.2.2(on FastEthernet0/0)
```

注意到所有网段都学习到了路由

任务 10：相关的查看命令。代码如下。

```
Router A#sh ip rip                //显示 RIP 状态
 Update: 30,   Expire: 180,   Holddown: 120
 Input-queue: 50
 Validate-update-source: Enable
 Neighbor List:

Router A#show ip rip protocol                    //显示协议细节
RIP is Active
update interval 30(s), Invalid interval 180(s)      //注意定时器的值
Holddown interval 120(s), Trigger interval 5(s)
Automatic network summarization: Enable
Network List:
  network 192.168.0.0
  network 192.168.1.0
Filter list:
Offset list:
Redistribute policy:
Interface send version and receive version:
Global version : default
  Interface                Send-version    Recv-version    Nbr_number
  Serial0/2                V1              V1 V2           2
  FastEthernet0/0          V1              V1 V2           0
Distance: 0 (default is 120):         //注意默认的管理距离
Maximum router count: 1024,         Current router count:4

Router A#show ip rip database        //显示 RIP 数据库
 192.168.0.0/24      directly connected    FastEthernet0/0
 192.168.1.0/24      directly connected    Serial0/2
 192.168.2.0/24      [120,1]   via 192.168.1.2 (on Serial0/2)    00:00:04      //收到 RIP 广播的时间
 192.168.3.0/24      [120,16]   via 192.168.1.2 holddown (on Serial0/2)    00:00:10
Router A#show ip router rip        //仅显示 RIP 学习到的路由
R        192.168.2.0/24        [120,1] via 192.168.1.2(on Serial0/2)
R        192.168.3.0/24        [120,2] via 192.168.1.2(on Serial0/2)
```

相关命令详解

在 DCNOS 中运行 RIP 的基本配置很简单，通常只需打开 RIP 开关，并且配置运行 RIP

的网段，即按 RIP 默认配置发送和接收 RIP 数据报。如果需要可以切换发送、接收 RIP 数据报的版本，允许/禁止发送、接收 RIP 数据报。见表 6-5。

表　6-5

<table>
<tr><th>命　　令</th><th>解　　释</th></tr>
<tr><td>router rip
no router rip</td><td>打开 RIP
本命令的 no 操作关闭 RIP</td></tr>
<tr><td>network <A.B.C.D/M | ifname>
no network <A.B.C.D/M | ifname></td><td>设定运行 RIP 的网段
本命令的 no 操作为删除运行 RIP 的网段</td></tr>
</table>

共同思考

1）为什么 B 没有配置 RIP 时，A 没有出现 RIP 路由？
2）如果不是连续的子网，会出现什么结果？
3）RIP 的广播周期是多少？

实训 4　路由器单区域 OSPF 协议的配置

实训目标

OSPF 协议具有快速收敛、无自环、区域划分、等价路由、路由分级、支持验证、组播发送等特性，在大规模网络中，OSPF 作为链路状态路由协议的代表应用非常广泛。通过本实训，我们可以掌握单区域 OSPF 的配置，从而更好地理解链路状态路由协议的工作过程。

实训设备

1）DCR2655/2659 路由器 2 台。
2）背靠背线 1 条。

实训拓扑

实训拓扑图如图 6-4 所示。

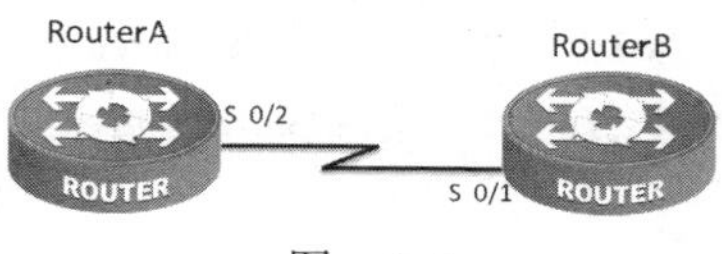

图　6-4

实训任务

IP 地址表，见表 6-6。

表 6-6

设　　备	S 0/1(DTE)	S0/2(DCE)	Loopback
Router A		192.168.1.1/24	10.10.10.1/24
Router B	192.168.1.2/24		10.10.11.1/24

任务 1：路由器环回接口的配置（其他接口配置请参见相关实训）。

```
路由器 A:
Router A_config#interface    loopback0                                      //设置 loop 口
Router A_config_l0#ip address 10.10.10.1 255.255.255.0
Router A_config_l0#ip ospf network point-to-point                          //还原 loop 口地址
路由器 B:
Router B#config
Router B_config#interface loopback0
Router B_config_l0#ip address   10.10.11.1 255.255.255.0
Router B_config_l0#ip ospf network point-to-point
```

任务 2：验证接口配置。代码如下。

```
Router A #sh interface loopback0
Loopback0 is up, line protocol is up
  Hardware is Loopback
  Interface address is 10.10.10.1/24
  MTU 1514 bytes, BW 8000000 kbit, DLY 500 usec
  Encapsulation LOOPBACK
Router B #sh interface loopback0
Loopback0 is up, line protocol is up
  Hardware is Loopback
  Interface address is 10.10.11.1/24
  MTU 1514 bytes, BW 8000000 kbit, DLY 500 usec
  Encapsulation LOOPBACK
```

任务 3：路由器的 OSPF 配置。代码如下。

```
A 的配置：
Router A_config#router ospf 1                    //启动 OSPF 进程，进程号为 1
Router A_config_ospf_1#network 10.10.10.0 255.255.255.0 area 0      //注意要写掩码和区域号
Router A_config_ospf_1#network 192.168.1.0 255.255.255.0 area 0
B 的配置：
Router B_config#router ospf 1
```

```
Router B_config_ospf_1#network 10.10.11.0 255.255.255.0 area 0
Router B_config_ospf_1#network 192.168.1.0 255.255.255.0 area 0
```

任务 4：查看路由表。代码如下。

```
路由器 A：
Router A#sh ip router
Codes: C - connected, S - static, R - RIP, B - BGP, BC - BGP connected
       D - DEIGRP, DEX - external DEIGRP, O - OSPF, OIA - OSPF inter area
       ON1 - OSPF NSSA external type 1, ON2 - OSPF NSSA external type 2
       OE1 - OSPF external type 1, OE2 - OSPF external type 2
       DHCP - DHCP type
VRF ID: 0
C      10.10.10.0/24      is directly connected, Loopback0
O      10.10.11.1/32      [110,1600] via 192.168.1.2(on Serial0/2)
                                     //注意到环回接口产生的是主机路由
C      192.168.1.0/24     is directly connected, Serial0/2
路由器 B：
Router B#show ip router
Codes: C - connected, S - static, R - RIP, B - BGP, BC - BGP connected
       D - DEIGRP, DEX - external DEIGRP, O - OSPF, OIA - OSPF inter area
       ON1 - OSPF NSSA external type 1, ON2 - OSPF NSSA external type 2
       OE1 - OSPF external type 1, OE2 - OSPF external type 2
       DHCP - DHCP type
VRF ID: 0
O      10.10.10.1/32     [110,1601] via 192.168.1.1(on Serial0/1)   //注意管理距离为 110
C      10.10.11.0/24      is directly connected, Loopback0
C      192.168.1.0/24     is directly connected, Serial0/1
```

任务 5：其他验证命令。代码如下。

```
Router B#show ip ospf 1          //显示该 OSPF 进程的信息
OSPF process: 1, Router ID: 10.10.11.1
Distance: intra-area 110,   inter-area 110,   external 150
SPF schedule delay 5 secs, Hold time between two SPFs 10 secs
SPFTV:0(0), TOs:6, SCHDs:6
All Rtrs support Demand-Circuit.
Number of areas is 1
AREA: 0
  Number of interface in this area is 2(UP: 1)
  Area authentication type:   None
  All Rtrs in this area support Demand-Circuit.
```

```
Router A#show ip ospf interface          //显示 OSPF 接口状态和类型
Loopback0 is up, line protocol is up
  Internet Address: 10.10.10.1/24
  Interface index: 9
  Nettype: Point-to-Point             //环回接口的网络类型为点对点
  OSPF process is 1,   AREA: 0, Router ID: 10.10.10.1
  Cost: 1, Transmit Delay is 1 sec, Priority 0
  Hello interval is 10, Dead timer is 40, Retransmit is 5
  OSPF INTF State is ILOOPBACK
  Neighbor Count is 0, Adjacent neighbor count is 0

Serial0/2 is up, line protocol is up
  Internet Address: 192.168.1.1/24
  Interface index: 3
  Nettype: Point-to-Point
  OSPF process is 1,   AREA: 0, Router ID: 10.10.10.1
  Cost: 1600, Transmit Delay is 1 sec, Priority 1
  Hello interval is 10, Dead timer is 40, Retransmit is 5
  OSPF INTF State is IPOINT_TO_POINT
  Neighbor Count is 1, Adjacent neighbor count is 1
    Adjacent with neighbor 192.168.1.2

Router A#sh ip ospf neighbor       //显示 OSPF 邻居
--------------------------------------------------------------------------------------------------------
                              OSPF process: 1
                                AREA: 0
Neighbor ID      Pri    State            DeadTime    Neighbor Addr    Interface
10.10.11.1       1      FULL/-           37          192.168.1.2      Serial0/2
```

任务 6：查看路由器 B 的路由表。代码如下。

```
Router B#show ip router
Codes: C - connected, S - static, R - RIP, B - BGP, BC - BGP connected
        D - BEIGRP, DEX - external BEIGRP, O - OSPF, OIA - OSPF inter area
        ON1 - OSPF NSSA external type 1, ON2 - OSPF NSSA external type 2
        OE1 - OSPF external type 1, OE2 - OSPF external type 2
        DHCP - DHCP type, L1 - IS-IS level-1, L2 - IS-IS level-2
VRF ID: 0
O       10.10.10.1/32            [110,1601] via 192.168.1.1(on Serial0/1)
```

```
C        10.10.11.0/24        is directly connected, Loopback0
C        192.168.1.0/24       is directly connected, Serial0/1
```

注意事项和排错

1）每个路由器的 OSPF 进程号可以不同，一个路由器可以有多个 OSPF 进程。
2）OSPF 是无类路由协议，一定要加掩码。
3）第一个区域必须是区域 0。
命令详解见表 6-7

表　6-7

命　　令	目　　的
ip ospf network point-to-point	还原 loop 口掩码
router ospf 1	启动 OSPF 进程，进程号为 1
network 10.10.10.0 255.255.255.0 area 0	注意要写掩码和区域号

共同思考

1）OSPF 与 RIP 有哪些区别？
2）环回接口有什么好处？

第 7 章　路由器接口设置实践

实训 1　路由器串口 PPP-PAP 配置

实训目标

今天小刘的领导为小刘分配了一个任务，为一个局端用户分配广域网链路接口，并针对广域网用户配置 PAP 方式的验证。在运营商网络中，一般使用串口实现广域网链路的互联，小刘先在公司使用两台路由器使用串口背靠背互联，模拟了运营商用户间的广域网链路 PAP 认证的配置。

实训设备

1）DCR2655/2659 路由器 2 台。

2）CR-V35FC 1/ CR-V35MT 1 各 1 条。

实训拓扑

实训拓扑图如图 7-1 所示。

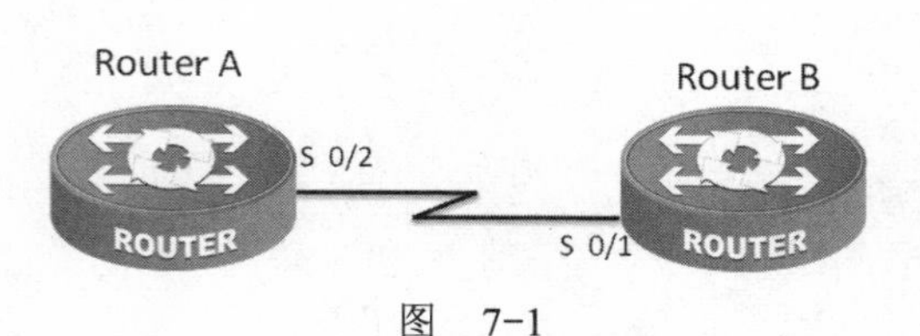

图　7-1

实训任务

任务 1：基本配置。

Router A		Router B	
接口	IP 地址	接口	IP 地址
S0/2 DCE	192.168.1.1	S0/1 DTE	192.168.1.2
账号	密码	账号	密码
Router A	digitalchinaA	Router B	digitalchinaB

任务 2：路由器 A 的配置。代码如下。

```
Router >enable                                                    //进入特权模式
Router #config                                                    //进入全局配置模式
Router _config#hostname Router A                                  //修改机器名
Router A_config#aaa authentication ppp test local
Router A_config#username Router B password digitalchinaB          //设置账号密码
Router A_config#interface s0/2                                    //进入接口模式
Router A_config_s0/2#ip address 192.168.1.1 255.255.255.0         //配置 IP 地址
Router A_config_s0/2#encapsulation PPP                            //封装 PPP 协议
Router A_config_s0/2#ppp authentication pap test                  //设置验证方式
Router A_config_s0/2#ppp pap sent-username Router A   password   digitalchinaA
                                                        //设置发送给对方验证的账号密码
Router-A_config_s0/2#physical-layer speed 64000                   //配置 DCE 时钟频率
Router-A_config_s0/2#no shutdown
Router-A_config_s0/2#^Z
```

任务 3：查看路由器 A 的配置。代码如下。

```
Router A#show interface s0/2                        //查看接口状态
Serial0/2 is up, line protocol is down              //对端没有设置，所以协议是 DOWN
 Mode=Sync DCE Speed=64000                          //查看 DCE
  DTR=UP,DSR=UP,RTS=UP,CTS=UP,DCD=UP
  MTU 1500 bytes, BW 64 kbit, DLY 2000 usec
  Interface address is 192.168.1.1/24               //查看 IP 地址
  Encapsulation PPP, loopback not set               //查看封装协议
  Keepalive set(10 sec)
  LCP   Starting configuration exchange
  PAP   Closed,   Message: 'none'
  IPCP Listening -- waiting for remote host to attempt open
        local IP address: 192.168.1.1    remote IP address: 0.0.0.0
  60 second input rate 0 bits/sec, 0 packets/sec!
  60 second output rate 0 bits/sec, 0 packets/sec!
    52 packets input, 1248 bytes, 4 unused_rx, 0 no buffer
    0 input errors, 0 CRC, 0 frame, 0 overrun, 0 ignored, 0 abort
    76 packets output, 1800 bytes, 8 unused_tx, 0 underruns
  error:
    0 clock, 0 grace
  PowerQUICC SCC specific errors:
    0 recv allocb mblk fail       0 recv no buffer
    0 transmitter queue full      0 transmitter hwqueue_full
```

任务 4：路由器 B 的配置。代码如下。

```
Router >enable                                                  //进入特权模式
Router #config                                                  //进入全局配置模式
Router _config#hostname Router B                                //修改机器名
Router B_config#aaa authentication ppp test local
Router B_config#username Router A password digitalchinaA        //设置账号密码
Router B_config#interface s0/1                                  //进入接口模式
Router B_config_s0/1#ip address 192.168.1.2 255.255.255.0       //配置 IP 地址
Router B_config_s0/1#encapsulation PPP                          //封装 PPP
Router B_config_s0/1#ppp authentication pap test                //设置验证方式
Router B_config_s0/1#ppp pap sent-username Router B password digitalchinaB
                                                   //设置发送给对方验证的账号密码
Router B_config_s0/1#shutdown
Router B_config_s0/1#no shutdown
Router B_config_s0/1#^Z                                         //按<ctrl + z>键进入特权模式
```

任务 5：查看路由器 B 的配置。代码如下。

```
Router B#show interface s0/1                                    //查看接口状态
Serial0/1 is up, line protocol is up                            //接口和协议都是 up
 Mode=Sync DTE                                                  //查看 DTE
  DTR=UP,DSR=UP,RTS=UP,CTS=UP,DCD=UP
  MTU 1500 bytes, BW 64 kbit, DLY 2000 usec
  Interface address is 192.168.1.2/24                           //查看 IP 地址
  Encapsulation PPP, loopback not set                           //查看封装协议
  Keepalive set(10 sec)
  LCP   Opened
  PAP   Opened,   Message: 'Welcome to Digital China Networks Limited Router'
  IPCP Opened
       local IP address: 192.168.1.2    remote IP address: 192.168.1.1
  60 second input rate 26 bits/sec, 0 packets/sec!
  60 second output rate 23 bits/sec, 0 packets/sec!
    15 packets input, 394 bytes, 5 unused_rx, 0 no buffer
    0 input errors, 0 CRC, 0 frame, 0 overrun, 0 ignored, 0 abort
    14 packets output, 370 bytes, 8 unused_tx, 0 underruns
  error:
    0 clock, 0 grace
  PowerQUICC SCC specific errors:
    0 recv allocb mblk fail       0 recv no buffer
    0 transmitter queue full      0 transmitter hwqueue_full
```

任务 6：检测连通性。代码如下。

```
Router A#ping 192.168.1.2
PING 192.168.1.2 (192.168.1.2): 56 data bytes
!!!!!
--- 192.168.1.2 ping statistics ---
5 packets transmitted, 5 packets received, 0% packet loss
round-trip min/avg/max = 20/20/20 ms

Router B#ping 192.168.1.1
PING 192.168.1.1 (192.168.1.1): 56 data bytes
!!!!!
--- 192.168.1.1 ping statistics ---
5 packets transmitted, 5 packets received, 0% packet loss
round-trip min/avg/max = 20/20/20 ms
```

注意事项和排错

1）账号密码一定要交叉对应，发送的账号密码要和对方账号数据库中的账号密码对应。

2）不要忘记配置 DCE 的时钟频率。

命令详解（见表 7-1）

ppp authentication

使用接口配置命令 ppp authentication 指定接口上使用 CHAP 或 PAP 的次序，使用 no ppp authentication 取消认证。

ppp authentication {**chap**|**ms-chap**|**pap**}[[*list-name*|**default**][**callin**]

no ppp authentication

表　7-1

命　令	目　的
chap	在串行接口上激活 CHAP
pap	在串行接口上激活 PAP
ms-chap	在串行接口上激活 MS-CHAP
list-name	（可选的）与 AAA/TACACS+一起使用，指定执行认证时使用的 TACACS＋方法列表名。如果没有指定列表名，系统将使用默认列表。使用命令 aaa authentication ppp 创建列表
default	（可选的）与 AAA/TACACS+一起使用。使用命令 aaa authentication ppp 创建默认列表
callin	（可选的）指定仅对收到的呼叫（calls）进行认证

共同思考

1）在什么环境下需要配置验证方式？

2）PAP 验证是否非常安全？

实训 2 路由器串口 PPP-CHAP 配置

实训目标

小刘在模拟运营商到用户广域网接入的实现中发现，使用 PPP-PAP 方式虽然可以通过用户名和密码限制合法用户的接入。但 PAP 的交互过程采用的是明文认证，这样的认证方式很容易被破解和攻击。小刘在和用户交流后，建议用户采用 PPP-CHAP 的方式对广域网链路进行加密认证。

实训设备

1）DCR2655/2659 路由器 2 台。
2）CR-V35FC 1/ CR-V35MT 1 各 1 条。

实训拓扑

实训拓扑图如图 7-2 所示。

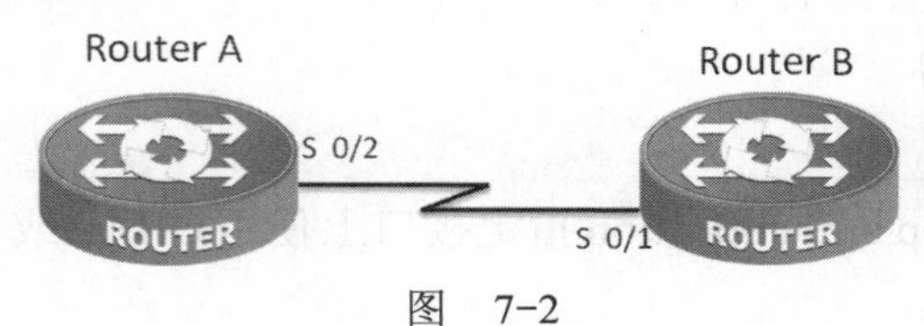

图 7-2

实训任务

任务 1：基本配置。代码如下。

Router A		Router B	
接口	IP 地址	接口	IP 地址
S0/2 DCE	192.168.1.1	S0/1 DTE	192.168.1.2
账号	密码	账号	密码
Router A	digitalchina	Router B	digitalchina

任务 2：路由器 A 的配置。代码如下。

```
Router >enable                                   //进入特权模式
Router #config                                   //进入全局配置模式
Router _config#hostname Router A                 //修改机器名
Router A_config#aaa authentication ppp test local
        //定义一个名为 test，使用本地数据进行验证的 aaa 验证方法
```

```
Router A_config#username Router B password digitalchina          //设置账号密码
Router A_config#interface s0/2                                   //进入接口模式
Router A_config_s0/2#ip address 192.168.1.1 255.255.255.0        //配置 IP 地址
Router A_config_s0/2#encapsulation PPP                           //封装 PPP
Router A_config_s0/2#ppp authentication chap test                //设置验证方式
Router A_config_s0/2#ppp chap hostname Router A                  //设置发送给对方验证的账号密码
Router-A_config_s0/2#physical-layer speed 64000                  //配置 DCE 时钟频率
Router-A_config_s0/2#no shutdown
Router-A_config_s0/2#^Z
```

任务 3：查看路由器 A 的配置。代码如下。

```
Router A#show interface s0/2                                     //查看接口状态
Serial0/2 is up, line protocol is down                           //对端没有设置，所以协议是 DOWN
 Mode=Sync DCE Speed=64000                                       //查看 DCE
  DTR=UP,DSR=UP,RTS=UP,CTS=UP,DCD=UP
  MTU 1500 bytes, BW 64 kbit, DLY 2000 usec
  Interface address is 192.168.1.1/24                            //查看 IP 地址
  Encapsulation PPP, loopback not set                            //查看封装协议
  Keepalive set(10 sec)
  LCP   Starting configuration exchange
  PAP   Closed,   Message: 'none'
  IPCP Listening -- waiting for remote host to attempt open
       local IP address: 192.168.1.1    remote IP address: 0.0.0.0
  60 second input rate 0 bits/sec, 0 packets/sec!
  60 second output rate 0 bits/sec, 0 packets/sec!
    52 packets input, 1248 bytes, 4 unused_rx, 0 no buffer
    0 input errors, 0 CRC, 0 frame, 0 overrun, 0 ignored, 0 abort
    76 packets output, 1800 bytes, 8 unused_tx, 0 underruns
  error:
    0 clock, 0 grace
  PowerQUICC SCC specific errors:
    0 recv allocb mblk fail        0 recv no buffer
    0 transmitter queue full       0 transmitter hwqueue_full
```

任务 4：路由器 B 的配置。代码如下。

```
Router >enable                                                   //进入特权模式
Router #config                                                   //进入全局配置模式
Router _config#hostname Router B                                 //修改机器名
Router B_config#aaa authentication ppp test local
Router B_config#username Router A password digitalchina          //设置账号密码
Router B_config#interface s0/1                                   //进入接口模式
```

```
Router B_config_s0/1#ip address 192.168.1.2 255.255.255.0      //配置 IP 地址
Router B_config_s0/1#encapsulation PPP                         //封装 PPP
Router B_config_s0/1#ppp authentication chap test              //设置验证方式
Router B_config_s0/1#ppp chap hostname Router B                //设置发送给对方验证的账号密码
Router B_config_s0/1#shutdown
Router B_config_s0/1#no shutdown
Router B_config_s0/1#^Z                                        //按<ctrl + z>键进入特权模式
```

任务 5：查看路由器 B 的配置。代码如下。

```
Router B#show interface s0/1                                   //查看接口状态
Serial0/1 is up, line protocol is up                           //接口和协议都是 up
 Mode=Sync DTE                                                 //查看 DTE
  DTR=UP,DSR=UP,RTS=UP,CTS=UP,DCD=UP
  MTU 1500 bytes, BW 64 kbit, DLY 2000 usec
  Interface address is 192.168.1.2/24                          //查看 IP 地址
  Encapsulation PPP, loopback not set                          //查看封装协议
  Keepalive set(10 sec)
  LCP    Opened
  PAP    Opened,    Message: 'Welcome to Digital China Networks Limited Router'
  IPCP Opened
        local IP address: 192.168.1.2    remote IP address: 192.168.1.1
  60 second input rate 26 bits/sec, 0 packets/sec!
  60 second output rate 23 bits/sec, 0 packets/sec!
    15 packets input, 394 bytes, 5 unused_rx, 0 no buffer
    0 input errors, 0 CRC, 0 frame, 0 overrun, 0 ignored, 0 abort
    14 packets output, 370 bytes, 8 unused_tx, 0 underruns
  error:
    0 clock, 0 grace
  PowerQUICC SCC specific errors:
    0 recv allocb mblk fail          0 recv no buffer
    0 transmitter queue full         0 transmitter hwqueue_full
```

任务 6：检测连通性。代码如下。

```
Router A#ping 192.168.1.2
PING 192.168.1.2 (192.168.1.2): 56 data bytes
!!!!!
--- 192.168.1.2 ping statistics ---
5 packets transmitted, 5 packets received, 0% packet loss
round-trip min/avg/max = 20/20/20 ms

Router B#ping 192.168.1.1
```

```
PING 192.168.1.1 (192.168.1.1): 56 data bytes
!!!!!
--- 192.168.1.1 ping statistics ---
5 packets transmitted, 5 packets received, 0% packet loss
round-trip min/avg/max = 20/20/20 ms
```

注意事项和排错

1）双方密码一定要一致，发送的账号要和对方账号数据库中的账号对应。
2）不要忘记配置 DCE 的时钟频率。
命令详解（见表 7-2）

表　7-2

命　　令	目　　的
ppp chap hostname Router B	设置发送给对方验证的账号
encapsulation PPP	封装 PPP
ppp authentication chap　test	设置验证方式

共同思考

1）CHAP 和 PAP 这两种验证有什么不同？
2）CHAP 验证是否非常安全？

第 8 章　路由器应用技术实践

实训 1　标准访问列表配置

实训目标

一天小张接到某用户打来的电话，反馈说内网中有一台 PCB，由于 PCB 上运行了重要的财务数据，用户不希望内网中其他网段的用户访问到这台 PCB。小张想到用户的这种应用可以使用访问列表来控制用户间数据互访。为了更好地解决用户的问题，小张在向用户答复前利用手边的路由器进行了标准访问列表的实验。

实训设备

1）DCR2655/2659 路由器 2 台。

2）CR-V35FC 1/ CR-V35MT 1 各 1 条，交叉双绞线 2 条。

3）PC2 台。

实训拓扑

实训拓扑图如图 8-1 所示。

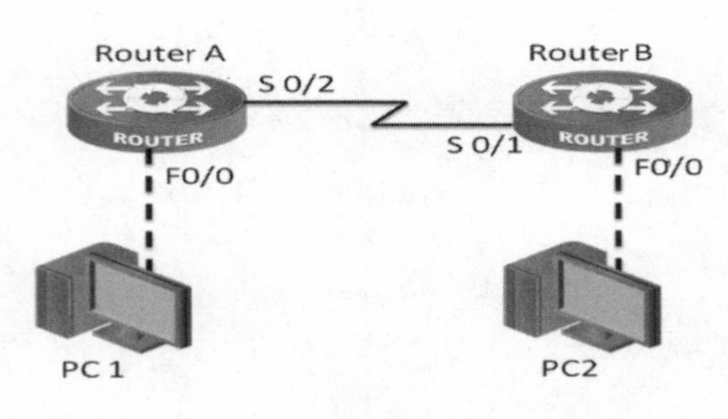

图　8-1

实训任务

任务 1：参照相关实训和表 8-1，配置所有接口的地址，并测试连通性。

表　8-1

主 机 名	F0/0	S0/1(DTE)	S0/2(DCE)	PC 地 址	默 认 网 关
Router A	192.168.0.1/24		192.168.1.1/24		
Router B	192.168.2.1/24	192.168.1.2/24			
PC1				192.168.0.2/24	192.168.0.1
PC2				192.168.2.2/24	192.168.2.1

```
Router A#ping 192.168.1.2
PING 192.168.1.2 (192.168.1.2): 56 data bytes
!!!!!
--- 192.168.1.2 ping statistics ---
5 packets transmitted, 5 packets received, 0% packet loss
round-trip min/avg/max = 20/28/30 ms
```

任务 2：参照相关实训，配置静态路由。代码如下。

```
Router A#show ip route
Codes: C - connected, S - static, R - RIP, B - BGP, BC - BGP connected
        D - BEIGRP, DEX - external BEIGRP, O - OSPF, OIA - OSPF inter area
        ON1 - OSPF NSSA external type 1, ON2 - OSPF NSSA external type 2
        OE1 - OSPF external type 1, OE2 - OSPF external type 2
        DHCP - DHCP type, L1 - IS-IS level-1, L2 - IS-IS level-2
VRF ID: 0
C       192.168.0.0/24          is directly connected, FastEthernet0/0
C       192.168.1.0/24          is directly connected, Serial0/2
S       192.168.2.0/24          [1,0] via 192.168.1.2(on Serial0/2)

Router B#show ip route
Codes: C - connected, S - static, R - RIP, B - BGP, BC - BGP connected
        D - BEIGRP, DEX - external BEIGRP, O - OSPF, OIA - OSPF inter area
        ON1 - OSPF NSSA external type 1, ON2 - OSPF NSSA external type 2
        OE1 - OSPF external type 1, OE2 - OSPF external type 2
        DHCP - DHCP type, L1 - IS-IS level-1, L2 - IS-IS level-2
VRF ID: 0
S       192.168.0.0/24          [1,0] via 192.168.1.1(on Serial0/1)
C       192.168.1.0/24          is directly connected, Serial0/1
C       192.168.2.0/24          is directly connected, FastEthernet0/0
```

任务 3：PC-A 能与 PC-B 通信，如图 8-2 所示。

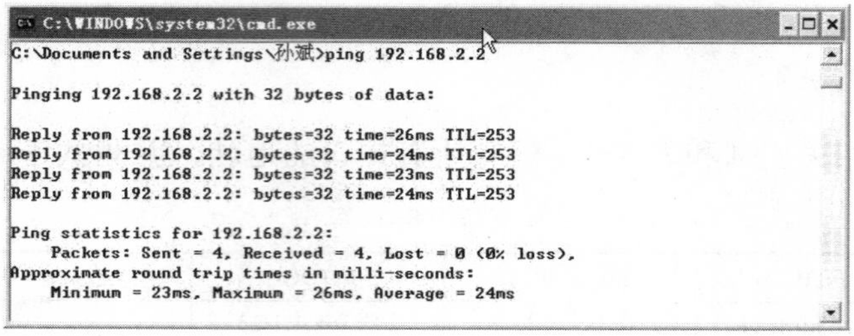

```
C:\WINDOWS\system32\cmd.exe
C:\Documents and Settings\孙斌>ping 192.168.2.2

Pinging 192.168.2.2 with 32 bytes of data:

Reply from 192.168.2.2: bytes=32 time=26ms TTL=253
Reply from 192.168.2.2: bytes=32 time=24ms TTL=253
Reply from 192.168.2.2: bytes=32 time=23ms TTL=253
Reply from 192.168.2.2: bytes=32 time=24ms TTL=253

Ping statistics for 192.168.2.2:
    Packets: Sent = 4, Received = 4, Lost = 0 (0% loss),
Approximate round trip times in milli-seconds:
    Minimum = 23ms, Maximum = 26ms, Average = 24ms
```

图 8-2

任务 4：配置访问控制列表禁止 PC1 所在的网段对 PC2 的访问。代码如下。

```
Router B#conf
Router B_config#ip access-list standard 1                    //定义标准的访问控制列表
Router B_config_std_nacl#deny 192.168.0.0 255.255.255.0      //基于源地址
Router B_config_std_nacl#permit any                          //因为有隐含的 DENY ANY
```

任务 5：将访问控制列表（ACL）绑定在相应的接口。代码如下。

```
Router B_config#int f0/0                          //进入到离目标最近的接口
Router B_config_f0/0#ip access-group 1 out        //绑定 ACL 1 在出口的方向
```

任务 6：验证。代码如下。

```
Router B#sh ip access-list
Standard IP access list 1
    deny     192.168.0.0 255.255.255.0
    permit any
```

任务 7：测试。如图 8-3 所示。

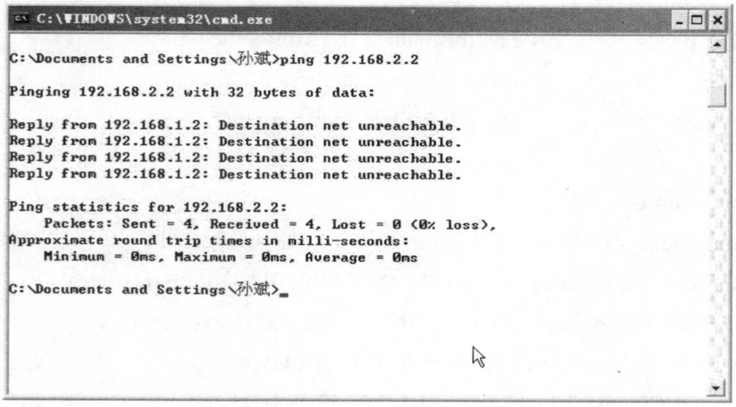

```
C:\WINDOWS\system32\cmd.exe
C:\Documents and Settings\孙斌>ping 192.168.2.2

Pinging 192.168.2.2 with 32 bytes of data:

Reply from 192.168.1.2: Destination net unreachable.
Reply from 192.168.1.2: Destination net unreachable.
Reply from 192.168.1.2: Destination net unreachable.
Reply from 192.168.1.2: Destination net unreachable.

Ping statistics for 192.168.2.2:
    Packets: Sent = 4, Received = 4, Lost = 0 (0% loss),
Approximate round trip times in milli-seconds:
    Minimum = 0ms, Maximum = 0ms, Average = 0ms

C:\Documents and Settings\孙斌>_
```

图 8-3

注意事项和排错

1）标准访问控制列表是基于源地址的。

2）每条访问控制列表都有隐含的拒绝。

3）标准访问控制列表一般绑定在离目标最近的接口。

4）注意方向，以该接口为参考点，IN 是流进的方向；OUT 是流出的方向。

相关命令详解（见表 8-2）

deny

表　8-2

protocol	协议名字或 IP 号。它可以是关键字 icmp、igmp、igrp、ip、ospf、tcp 或 udp，也可以是表 IP 号的 0～255 的一个整数。为了匹配任何 Internet 协议（包括 ICMP、TCP 和 UDP）使用关键字 IP。某些协议允许进一步限定
source	源网络或主机号。有两种方法指定源：32 位二进制数，用四个点隔开的十进制数表示。使用关键字 any 作为 0.0.0.0 0.0.0.0 的源和源掩码缩写
source-mask	源地址网络掩码。使用关键字 any 作为 0.0.0.0 0.0.0.0 的源和源掩码缩写
destination	目标网络或主机号。有两种方法指定： ① 使用四个点隔开的十进制数表示的 32 位二进制数 ② 使用关键字 any 作为 0.0.0.0 0.0.0.0 的目标和目标掩码的缩写
destination-mask	目标地址网络掩码。使用关键字 any 作为 0.0.0.0 0.0.0.0 的目标地址和目标地址掩码缩写
precedence	（可选）包可以由优先级过滤，用 0～7 的数字指定
tos tos	（可选）数据包可以使用服务层过滤。使用数字 0～15 指定
icmp-type	（可选）ICMP 包可由 ICMP 报文类型过滤。类型是数字 0～255
igmp-type	（可选）IGMP 包可由 IGMP 报文类型或报文名过滤。 类型是 0～15 的数字
operator	（可选）比较源或目标端口。操作包括 lt（小于），gt（大于），eq（等于），neq（不等于）。如果操作符放在 source 和 source-mask 之后，那么它必须匹配这个源端口。如果操作符放在 destination 和 destination-mask 之后，那么它必须匹配目标端口
port	（可选）TCP 或 UDP 端口的十进制数字或名称。端口号是一个 0～65535 的数字。TCP 端口名列在“使用方针”部分。当过滤 TCP 时，可以只使用 TCP 端口名称。UDP 端口名称也列在“使用说明”部分。当过滤 TCP 时，只可使用 TCP 端口名。当过滤 UDP 时，只可使用 UDP 端口名
established	（可选）只对 TCP，表示一个已建立的连接。如果 TCP 数据报 ACK 或 RST 位设置时，出现匹配。非匹配的情况是初始化 TCP 数据报，以形成一个连接
log	（可选）可以进行日志记录

可以使用访问列表控制包在接口上的传输，控制虚拟终端线路访问以及限制路由选择更新的内容。在匹配发生以后停止检查扩展的访问表。分段 IP 包，而不是初始段，立即由任何扩展的 IP 访问表接收。扩展的访问表用于控制访问虚拟终端线路或限制路由选择更新的内容，不必匹配 TCP 源端口、服务值的类型或包的优先权。

ip access-group

参数，见表 8-3。

表　8-3

access-list-name	访问表名。这是一个最长为 20 个字符的字符串
in	在入接口时使用访问列表
out	在出接口时使用访问列表

访问列表既可用在出接口也可用在入接口。对于标准的入口访问列表，在接收到包之

后，对照访问列表检查包的源地址。对于扩展的访问列表，该路由器也检查目标地址。如果访问列表允许该地址，那么软件继续处理该包。如果访问列表不允许该地址，该软件放弃包并返回一个 ICMP 主机不可到达报文。

对于标准的出口访问列表，在接收和路由一个包到控制接口以后，软件对照访问列表检查包的源地址。对于扩展的访问列表，路由器还检查接收端地址。如果访问列表允许该软件就传送这个包。如果访问列表不允许该地址，软件放弃这个包并返回一个 ICMP 主机不可达报文。

如果指定的访问列表不存在，所有的包允许通过。

ip access-list

参数，见表 8-4。

表 8-4

standard	指定为标准访问列表
extended	指定为扩展访问列表
name	访问列表名。这是一个最长 20 的字符串

使用此命令将进入 IP 访问列表配置模式，在 IP 访问列表配置模式中，可以用 deny 或 permit 命令来配置访问规则。见表 8-5。

permit

表 8-5

protocol	协议名字或 IP 号。它可以是关键字 icmp、igmp、igrp、ip、ospf、tcp 或 udp，也可以是表 IP 号的 0～255 的一个整数。为了匹配任何 Internet 协议（包括 ICMP、TCP 和 UDP）使用关键字 IP。某些协议允许进一步限定
source	源网络或主机号。有两种方法指定源：32 位二进制数，用四个点隔开的十进制数表示。使用关键字 any 作为 0.0.0.0 0.0.0.0 的源和源掩码缩写
source-mask	源地址网络掩码。使用关键字 any 作为 0.0.0.0 0.0.0.0 的源和源掩码缩写
destination	目标网络或主机号。有两种方法指定： ① 使用四个点隔开的十进制数表示的 32 位二进制数 ② 使用关键字 any 作为 0.0.0.0 0.0.0.0 的目标和目标掩码的缩写
destination-mask	目标地址网络掩码。使用关键字 any 作为 0.0.0.0 0.0.0.0 的目标地址和目标地址掩码缩写
precedence *precedence*	（可选）包可以由优先级过滤，用 0～7 的数字指定
tos *tos*	（可选）数据包可以使用服务层过滤。使用数字 0～15 指定
icmp-type	（可选）ICMP 包可由 ICMP 报文类型过滤。类型是数字 0～255
igmp-type	（可选）IGMP 包可由 IGMP 报文类型或报文名过滤。 类型是 0～15 的数字
operator	（可选）比较源或目标端口。操作包括 lt（小于），gt（大于），eq（等于），neq（不等于）。如果操作符放在 source 和 source-mask 之后，那么它必须匹配这个源端口。如果操作符放在 destination 和 destination-mask 之后，那么它必须匹配目标端口
port	（可选）TCP 或 UDP 端口的十进制数字或名称。端口号是一个 0~65535 的数字。TCP 端口名列在“使用方针”部分。当过滤 TCP 时，可以只使用 TCP 端口名称。UDP 端口名称也列在“使用说明”部分。当过滤 TCP 时，只可使用 TCP 端口名。当过滤 UDP 时， 只可使用 UDP 端口名
established	（可选）只对 TCP，表示一个已建立的连接。如果 TCP 数据报 ACK 或 RST 位设置时，出现匹配。非匹配的情况是初始化 TCP 数据报，以形成一个连接
log	（可选）可以进行日志记录

可以使用访问列表控制包在接口上的传输，控制虚拟终端线路访问以及限制路由选择更新的内容。在匹配发生以后停止检查扩展的访问列表。

分段 IP 包，而不是初始段，立即由任何扩展的 IP 访问列表接收。扩展的访问列表用于控制访问虚拟终端线路或限制路由选择更新的内容，不必匹配 TCP 源端口、服务值的类型或包的优先权。

共同思考

1）为什么访问控制列表最后要加一条允许？

2）除了绑定在 F0/0 以外，在现在的环境中还能绑定在哪个接口上？什么方向？

实训 2　扩展访问列表配置

实训目标

小刘在向用户回访网络运行状态的过程中，用户提到了一个新的需求。现在网络中有一台 PCB，用户不希望内网用户可以使用 Telnet 方式登录到这台 PC。同时，内网中还有一台网管服务器，网管服务器使用 Ping 的方式检验内网 PC 是否在线。

小刘总结了用户的需求，针对 PCB 来说，一方面禁止内网用户针对 PCB 的 Telnet，另一方面需要允许 PCB 响应 Ping 的请求，这种应用场景因为涉及具体的应用，就需要使用 ACL 访问控制列表限制具体的应用端口了。

实训设备

1）DCR2655/2659 路由器 2 台。

2）CR-V35FC 1/ CR-V35MT 1 各 1 条，交叉双绞线 2 条。

3）PC2 台。

实训拓扑

实训拓扑图如图 8-4 所示。

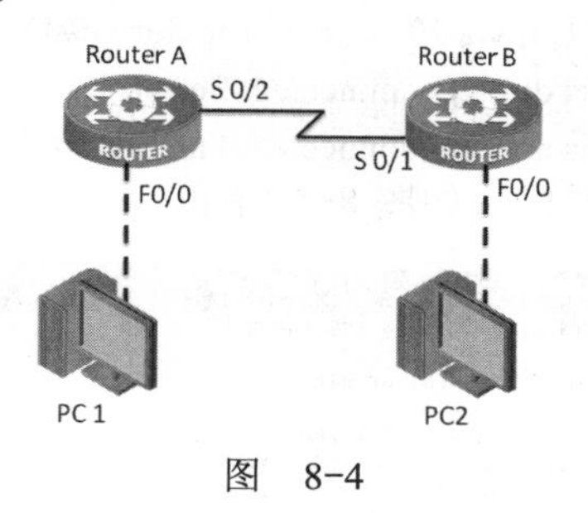

图　8-4

实训任务

任务 1：参照相关实训和表 8-6，配置所有接口的地址，并测试连通性。

表 8-6

主 机 名	F0/0	S0/1(DTE)	S0/2(DCE)	PC 地 址	默 认 网 关
Router A	192.168.0.1/24		192.168.1.1/24		
Router B	192.168.2.1/24	192.168.1.2/24			
PC1				192.168.0.2/24	192.168.0.1
PC2				192.168.2.2/24	192.168.2.1

```
Router-A#ping 192.168.1.2
PING 192.168.1.2 (192.168.1.2): 56 data bytes
!!!!!
--- 192.168.1.2 ping statistics ---
5 packets transmitted, 5 packets received, 0% packet loss
round-trip min/avg/max = 20/28/30 ms
```

任务 2：参照相关实训，配置静态路由。代码如下。

```
Router A#show ip router
Codes: C - connected, S - static, R - RIP, B - BGP, BC - BGP connected
        D - BEIGRP, DEX - external BEIGRP, O - OSPF, OIA - OSPF inter area
        ON1 - OSPF NSSA external type 1, ON2 - OSPF NSSA external type 2
        OE1 - OSPF external type 1, OE2 - OSPF external type 2
        DHCP - DHCP type, L1 - IS-IS level-1, L2 - IS-IS level-2
VRF ID: 0
C       192.168.0.0/24        is directly connected, FastEthernet0/0
C       192.168.1.0/24        is directly connected, Serial0/2
S       192.168.2.0/24        [1,0] via 192.168.1.2(on Serial0/2)
```

```
Router B#show ip router
Codes: C - connected, S - static, R - RIP, B - BGP, BC - BGP connected
        D - BEIGRP, DEX - external BEIGRP, O - OSPF, OIA - OSPF inter area
        ON1 - OSPF NSSA external type 1, ON2 - OSPF NSSA external type 2
        OE1 - OSPF external type 1, OE2 - OSPF external type 2
        DHCP - DHCP type, L1 - IS-IS level-1, L2 - IS-IS level-2
VRF ID: 0
S       192.168.0.0/24        [1,0] via 192.168.1.1(on Serial0/1)
C       192.168.1.0/24        is directly connected, Serial0/1
C       192.168.2.0/24        is directly connected, FastEthernet0/0
```

任务 3：PC-A 能与 PC-B 通信。如图 8-5 所示。

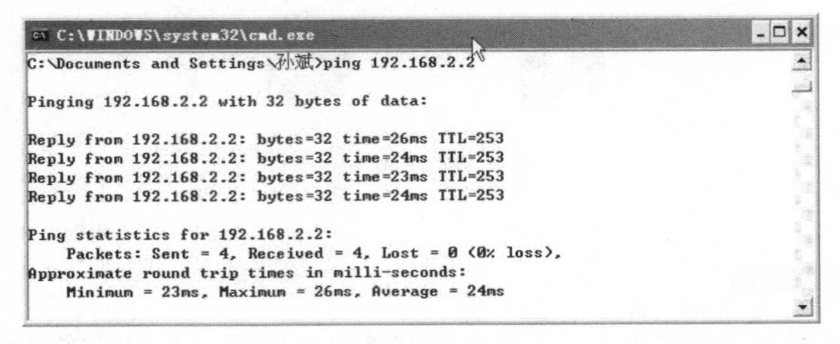

图 8-5

任务 4：在路由器 A 上设置访问列表。代码如下。

```
Router A#conf
Router A_config#ip access-list extended 192                        //定义扩展访问列表
Router A_config_ext_nacl#deny tcp 192.168.0.0 255.255.255.0 192.168.2.2 255.255.255.0 eq 23
                                                                    //设置扩展访问列表，拒绝 Telnet
Router A_config_ext_nacl#permit icmp any any                       //允许 Ping
Router A_config_ext_nacl#exit
Router A_config#int f0/0                                           //进入离源比较近的接口
Router A_config_f0/0#ip access-group 192 in                        //绑定访问列表在 IN 的方向
```

任务 5：查看访问列表。代码如下。

```
Router-A#sh ip access-list
Extended IP access list 192
    deny    tcp 192.168.0.0 255.255.255.0 192.168.2.2 255.255.255.0 eq telnet
    permit icmp any any
```

任务 6：验证。代码如下。

```
C:\Documents and Settings\Administrator>ping 192.168.2.2
Pinging 192.168.2.2 with 32 bytes of data:
Reply from 192.168.2.2: bytes=32 time=22ms TTL=62
Reply from 192.168.2.2: bytes=32 time=21ms TTL=62
Reply from 192.168.2.2: bytes=32 time=21ms TTL=62
Reply from 192.168.2.2: bytes=32 time=21ms TTL=62
Ping statistics for 192.168.2.2:
    Packets: Sent = 4, Received = 4, Lost = 0 (0% loss),
Approximate round trip times in milli-seconds:
    Minimum = 21ms, Maximum = 22ms, Average = 21ms

C:\Documents and Settings\Administrator>telnet 192.168.2.2
正在连接到 192.168.2.2...不能打开到主机的连接，在端口 23：连接失败
```

注意事项和排错

1）扩展访问列表通常放在离源比较近的地方。

2）扩展访问列表可以基于源、目标 IP、协议、端口号等条件过滤。

3）命令比较长，可以通过？的方式查看帮助。

4）注意隐含的 DENY。

共同思考

1）为什么扩展访问列表通常放在离源比较近的接口处？

2）访问列表中主机还有什么表示方法？

3）为什么访问列表中至少要有一条允许？

实训 3　路由器 NAT 实训

实训目标

小宋明天要去现场为用户架设一台出口路由器，想到出口路由器做为网络的出口设备，要为内网用户提供 NAT（网络地址转换）功能。就在本地先使用路由器进行了 NAT 功能的配置实验。

实训设备

1）DCR2655/2659 路由器 2 台。

2）CR-V35FC 1/ CR-V35MT 1 各 1 条，交叉双绞线 2 条。

3）PC2 台。

实训拓扑

实训拓扑图如图 8-6 所示。

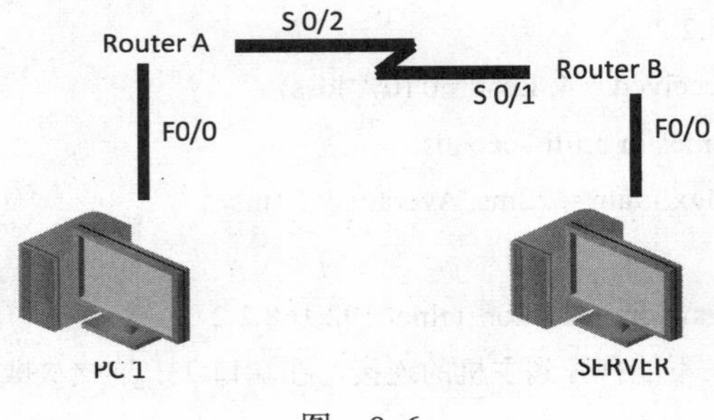

图　8-6

实训任务

任务 1：按相关实训和表 8-7 将接口地址和 PC 地址配置好，并且做连通性测试。

表　8-7

主　机　名	S0/1	F0/0	IP　地　址	网　　关
Router A	192.168.1.1/24	192.168.0.1/24		
Router B	192.168.1.2/24	192.168.2.1/24		
PC 1			192.168.0.3/24	192.168.0.1
SERVER			192.168.2.2/24	192.168.2.1

内部的 PC 需要访问外部的服务器。

假设在路由器 A 上做地址转换，将 192.168.0.0/24 转换成 192.168.1.10～192.168.1.20 之间的地址，并且做端口的地址复用。

任务 2：配置路由器 A 的 NAT。代码如下。

```
Router A#conf
Router A_config#ip access-list standard 1                  //定义访问控制列表
Router A_config_std_nacl#permit 192.168.0.0 255.255.255.0         //定义允许转换的源地址范围
Router A_config_std_nacl#exit
Router A_config#ip nat pool overld 192.168.1.10 192.168.1.20 255.255.255.0
                                                   //定义名为 overld 的转换地址池
Router A_config#ip nat inside source list 1 pool overld overload
                    //配置将 ACL 允许的源地址转换成 overld 中的地址，并且做 PAT 的地址复用
Router A_config#int f0/0
Router A_config_f0/0#ip nat inside                          //定义 F0/0 为内部接口
Router A_config_f0/0#int s0/1
Router A_config_s1/1#ip nat outside                         //定义 S0/1 为外部接口
Router A_config_s1/1#exit
Router A_config#ip route 0.0.0.0 0.0.0.0 192.168.1.2        //配置路由器 A 的默认路由
```

任务 3：查看路由器 B 的路由表。代码如下。

```
Router B#sh ip router
Codes: C - connected, S - static, R - RIP, B - BGP, BC - BGP connected
       D - DEIGRP, DEX - external DEIGRP, O - OSPF, OIA - OSPF inter area
       ON1 - OSPF NSSA external type 1, ON2 - OSPF NSSA external type 2
       OE1 - OSPF external type 1, OE2 - OSPF external type 2
       DHCP - DHCP type
VRF ID: 0
//注意：并没有到 192.168.0.0 的路由
```

任务 4：测试。如图 8-7 所示。

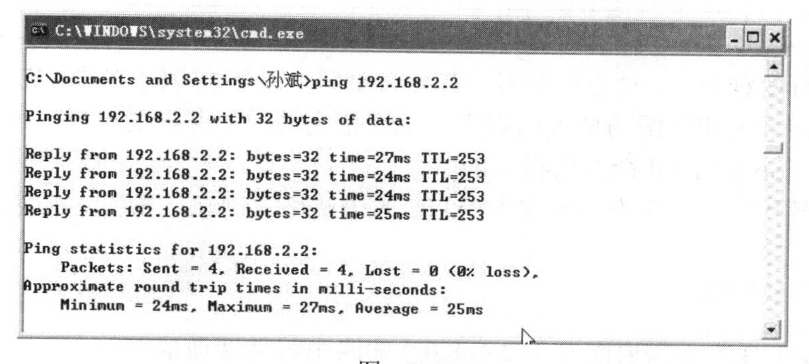

图　8-7

任务 5：查看地址转换表。代码如下。

```
Router A#sh ip nat translations
Pro. Dir     Inside local          Inside global         Outside local        Outside global

ICMP OUT 192.168.0.3:512   192.168.1.10:12512   192.168.1.2:12512   192.168.1.2:12512
C        192.168.1.0/24        is directly connected, Serial0/1
C        192.168.2.0/24        is directly connected, FastEthernet0/0
```

注意事项和排错

1）注意转换的方向和接口。

2）注意地址池、ACL 的名称。

3）需要配置 A 的默认路由。

命令详解（见表 8-8）

ip nat inside source

参数

表 8-8

参　数	参数说明
List access-list-name	IP 访问列表的名字。源地址符合访问列表的报文将被用地址池中的全局地址来翻译
pool name	地址池的名字，从这个池中动态地分配全局 IP 地址
interface type number	指定网络接口
overload	（可选）使路由器对多个本地地址使用一个全局的地址。当 OVERLOAD 被设置后，相同或者不同主机的多个会话将用 TCP 或 UDP 端口号来区分
static local-ip	建立一条独立的静态地址翻译；参数为分配给内部网主机的本地 IP 地址。这个地址可以自由选择，或从 RFC 1918 中分配
local-port	设置本地 TCP/UDP 端口号，范围为 1～65535
static global-ip	建立一条独立的静态地址翻译；这个参数为内部主机建立一个外部的网络可以唯一访问的 IP 地址
global-port	设置全局 TCP/UDP 端口号，范围为 1～65535
tcp	设置 TCP 端口翻译
udp	设置 UDP 端口翻译
network local-network	设置本地网段翻译
global-network	设置全局网段翻译
mask	设置网段翻译的网络掩码

这个命令有两种形式：动态的和静态的地址翻译。带有访问列表的格式建立动态翻译。来自和标准访问列表相匹配的地址的报文，将用指定的池中分配的全局地址来进行地址翻译，这个池是用 ip nat pool 命令所指定的。

可以作为替代方法，带有关键字 STATIC 的语法格式创建一条独立的静态地址翻译。

共同思考

1）为什么在 B 上不需要配置 192.168.0.0 的路由就能够通信？

2）为什么在 A 上需要配置默认路由？

3）如果外部接口地址是通过拨号动态获得，那么该如何配置？

实训 4　IPSec VPN（IKE）的配置

实训目标

今天有一个用户由于业务开展，需要在另一个城市设立办事处，为了使总部和办事处间使用同一平台办公。用户向运营商提出了在总部和办事处间建立 VPN 连接的请求。小李在接到用户的请求后，与用户沟通确定总部和办事处间使用公网链路互联，传输数据时有数据加密的需求。

基于上面的需求，小李认为使用 IPSce VPN 的方式可以更好地实现用户的需求。小李使用本地的两台路由器模拟了用户的应用环境，开始了 IPSec VPN 的配置。

实训设备

1）DCR2655/2659 路由器 2 台。
2）CR-V35FC 1/ CR-V35MT 1 各 1 条，交叉双绞线 2 条。
3）PC2 台。

实训拓扑

实训拓扑图如图 8-8 所示。

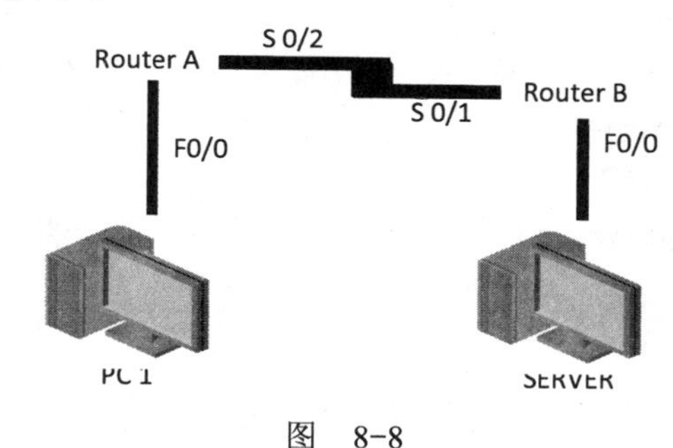

图　8-8

实训任务

任务 1：参照相关实训和表 8-9，配置所有接口的地址，并测试连通性。

表　8-9

主　机　名	F0/0	S0/1(DTE)	S0/2(DCE)	PC　地　址	默 认 网 关
Router A	192.168.0.1/24		192.168.1.1/24		
Router B	192.168.2.1/24	192.168.1.2/24			
PC1				192.168.0.2/24	192.168.0.1
Server				192.168.2.2/24	192.168.2.1

任务 2：路由器 A 的配置。代码如下。

```
Router A#conf
Router A_config#ip access-list extended 101          //确定要经过 VPN 保护的数据流
Router A_config_ext_nacl#permit ip 192.168.0.0 255.255.255.0 192.168.2.0 255.255.255.0
Router A_config_ext_nacl#exit
Router A_config#ip route 0.0.0.0 0.0.0.0 192.168.1.2                //配置静态路由
Router A_config#crypt isakmp policy 10                              //配置 IKE 策略
Router A_config_isakmp#authentication pre-share                     //设置认证方式
Router A_config_isakmp#encryption des                               //设置加密方式
Router A_config_isakmp#hash md5                                     //设置数字签名算法
Router A_config_isakmp#group 1                                      //设置 DH 方式
Router A_config_isakmp#lifetime 86400                               //设置生存期
Router A_config_isakmp#exit
Router A _config#crypto isakmp key digital 192.168.1.2              //设置共用密钥
Router A _config#crypto ipsec transform-set one                //设置变换集
Router A _config_crypto_trans#transform-type esp-des esp-md5-hmac   //ESP 加密和验证
Router A _config_crypto_trans#mode tunnel                           //设置为隧道模式
Router A _config_crypto_trans#exit
Router A _config#crypto map my 10 ipsec-isakmp                      //配置 IPSec 加密映射
Router A _config_crypto_map#set transform-set one                   //关联变换集
Router A _config_crypto_map#set peer 192.168.1.2               //设置对等体地址
Router A _config_crypto_map#match address 101                       //关联需要加密的数据流
Router A _config_crypto_map#exit
Router A _config#interface s0/2                                          //进入 VPN 的接口
Router A _config_s0/2#crypto map my                            //绑定 IPSec 加密映射
Router A _config_s0/2#^Z
```

任务 3：查看配置。代码如下。

```
Router A#sh crypto isakmp policy                        //查看 IKE 策略
Protection suite of priority 10
        encryption algorithm:     DES   - Data Encryption Standard (56 bit keys).
        hash algorithm:            Message Digest 5
        authentication method:    Pre-Shared Key
        Diffie-Hellman group:      #1 (768 bit)
        lifetime:                  86400 seconds
Default protection suite
        encryption algorithm:     DES   - Data Encryption Standard (56 bit keys).
        hash algorithm:            Secure Hash Standard
        authentication method:    Pre-Shared Key
        Diffie-Hellman group:      #1 (768 bit)
```

```
        lifetime:                      86400 seconds

Router A#sh crypto isakmp sa                          //查看 IKE 安全关联（没有建立，为空）

Router A#sh crypto map                                //查看 IPSec 映射
Crypto Map my 10 ipsec-isakmp
        Extended IP access list 101
          permit ip 192.168.0.0 255.255.255.0 192.168.2.0 255.255.255.0
        peer = 192.168.1.2
        PFS (Y/N): N
        Security association lifetime: 4608000 kilobytes/3600 seconds
        Transform sets={ one,}

Router-A#sh crypto ipsec sa                           //查看 IPSec 关联
Interface: Serial0/2
Crypto map name:my ,   local addr. 192.168.1.1
local    ident (addr/mask/prot/port): (192.168.0.0/255.255.255.0/0/0)
  remote ident (addr/mask/prot/port): (192.168.2.0/255.255.255.0/0/0)
  local crypto endpt.: 192.168.1.1,   remote crypto endpt.: 192.168.1.2

Router-A#sh crypto ipsec transform-set                //查看变换集
Transform set one: { esp-des esp-md5-hmac }
        will negotiate ={ Tunnel }
```

任务 4：路由器 B 的配置。代码如下。

```
Router B>ena
Router B#conf
Router B_config#ip access-list extended 101
Router B_config_ext_nacl#permit ip 192.168.2.0 255.255.255.0 192.168.0.0 255.255.255.0
Router B_config_ext_nacl#exit
Router B_config#ip route 192.168.0.0 255.255.255.0 192.168.1.1
Router B_config#crypto isakmp policy 10                  //注意与 A 要一致
Router B_config_isakmp#authentication pre-share
Router B_config_isakmp#hash md5
Router B_config_isakmp#encryption des
Router B_config_isakmp#group 1
Router B_config_isakmp#lifetime 86400
Router B_config_isakmp#exit
Router B_config#crypto isakmp key digital 192.168.1.1    //注意与 A 要一致
Router B_config#crypto ipsec transform-set one
```

```
Router B_config_crypto_trans#transform-type esp-des esp-md5-hmac     //注意与 A 要一致
Router B_config_crypto_trans#mode tunnel
Router B_config_crypto_trans#exit
Router B_config#crypto map my 10 ipsec-isakmp                        //注意与 A 要一致
Router B_config_crypto_map#set transform-set one
Router B_config_crypto_map#set peer 192.168.1.1
Router B_config_crypto_map#match address 101
Router B_config_crypto_map#exit
Router B_config#int s1/0
Router B_config_s1/0#crypto map my
Router B_config_s1/0#^Z
```

任务 5：查看配置。代码如下。

```
Router-B#sh crypto isakmp policy
Protection suite of priority 10
        encryption algorithm:      DES    - Data Encryption Standard (56 bit keys).
        hash algorithm:            Message Digest 5
        authentication method:     Pre-Shared Key
        Diffie-Hellman group:      #1 (768 bit)
        lifetime:                  86400 seconds
Default protection suite
        encryption algorithm:      DES    - Data Encryption Standard (56 bit keys).
        hash algorithm:            Secure Hash Standard
        authentication method:     Pre-Shared Key
        Diffie-Hellman group:      #1 (768 bit)
        lifetime:                  86400 secondsRouter-B#sh crypto isakmp sa

Router-B#sh crypto ipsec sa
Interface: Serial0/1
Crypto map name:my ,   local addr. 192.168.1.2
local   ident (addr/mask/prot/port): (192.168.2.0/255.255.255.0/0/0)
  remote ident (addr/mask/prot/port): (192.168.0.0/255.255.255.0/0/0)
  local crypto endpt.: 192.168.1.2,   remote crypto endpt.: 192.168.1.1

Router-B#sh crypto ipsec transform-set
Transform set one: { esp-des esp-md5-hmac }
        will negotiate ={ Tunnel }
Router-B#sh crypto map
Crypto Map my 10 ipsec-isakmp
        Extended IP access list 101
```

```
  permit ip 192.168.2.0 255.255.255.0 192.168.0.0 255.255.255.0
  peer = 192.168.1.1
  PFS (Y/N): N
  Security association lifetime: 4608000 kilobytes/3600 seconds
  Transform sets={ one,}
```

任务 6：测试。如图 8-9 所示。

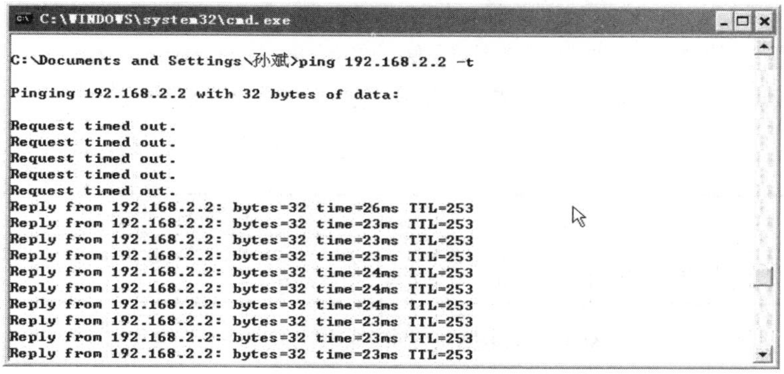

```
C:\WINDOWS\system32\cmd.exe

C:\Documents and Settings\孙斌>ping 192.168.2.2 -t

Pinging 192.168.2.2 with 32 bytes of data:

Request timed out.
Request timed out.
Request timed out.
Request timed out.
Request timed out.
Reply from 192.168.2.2: bytes=32 time=26ms TTL=253
Reply from 192.168.2.2: bytes=32 time=23ms TTL=253
Reply from 192.168.2.2: bytes=32 time=23ms TTL=253
Reply from 192.168.2.2: bytes=32 time=23ms TTL=253
Reply from 192.168.2.2: bytes=32 time=24ms TTL=253
Reply from 192.168.2.2: bytes=32 time=24ms TTL=253
Reply from 192.168.2.2: bytes=32 time=24ms TTL=253
Reply from 192.168.2.2: bytes=32 time=23ms TTL=253
Reply from 192.168.2.2: bytes=32 time=23ms TTL=253
Reply from 192.168.2.2: bytes=32 time=23ms TTL=253
```

图　8-9

再次查看安全关联：

```
Router-B#sh crypto isakmp sa
    dst            src            state            state-id     conn
192.168.1.1    192.168.1.2    <R>Q_SA_SETUP        2         2 my 10
192.168.1.1    192.168.1.2    <R>M_SA_SETUP        1         2 my 10

Router-B#sh crypto ipsec sa
Interface: Serial0/1
Crypto map name:my ,   local addr. 192.168.1.2
local   ident (addr/mask/prot/port): (192.168.2.0/255.255.255.0/0/0)
 remote ident (addr/mask/prot/port): (192.168.0.0/255.255.255.0/0/0)
 local crypto endpt.: 192.168.1.2,   remote crypto endpt.: 192.168.1.1
 inbound esp sas:
   spi:0x7f73737b(2138272635)
     transform:   esp-des   esp-md5-hmac
     in use settings ={ Tunnel }
     sa timing: remaining key lifetime (k/sec):   (4607998/3495)
outbound esp sas:
   spi:0x2cb62d75(750136693)
     transform:   esp-des   esp-md5-hmac
     in use settings ={ Tunnel }
     sa timing: remaining key lifetime (k/sec):   (4607998/3495)
```

注意事项和排错

1）注意两端参数要一致。

2）ACL 的作用是确定哪些数据需要经过 VPN。

命令详解（见表 8-10）

transform-type

加密变换配置状态下，要设置变换类型，使用 transform-type 命令。

```
transform-type    transform1 [transform2[transform3]]
```

表 8-10

参数	参数说明
transform1/ transform2/ transform3	可以指定 3 个以下的变换。这些变换定义了 IPSec 安全协议和算法。可接受的变换值在“使用说明”中详细阐述

默认：默认的变换类型为 ESP-DES（ESP 采用 DES 加密算法）。

命令模式：加密变换配置状态。

使用说明：

变换集合可以指定一个或两个 IPSec 安全协议（或 ESP，或 AH，或两者都有），并且指定和选定的安全协议一起使用哪种算法。ESP 和 AH IPSec 安全协议在“IPSec 协议：封装安全协议和校验头”一节中做了详细阐述。

变换集合的定义可以指定 1～3 个变换——每个变换代表一个 IPSec 安全协议（ESP 或 AH）和想要使用的算法的组合。当 IPSec 安全联盟协商时使用了某一变换集合，整个变换集合（协议、算法和其他设置的组合）必须和对端的一个变换集合相匹配。

在一个变换集合中，可以指定 AH 协议、ESP 或两者都指定。如果在变换集合中指定了一个 ESP，那么可以只定义 ESP 加密变换，也可以 ESP 加密变换和 ESP 验证变换两者都定义。

表 8-11 中显示了可行的变换组合。

表 8-11

为变换集合选择变换：可行的变换组合					
AH 变换中选择一种		ESP 加密变换中选择一种		ESP 验证变换中选择一种	
变换	描述	变换	描述	变换	描述
ah-md5-hmac	带 MD5（HMAC 变量）的 AH 验证算法	esp-des	采用 DES 的 ESP 加密算法	esp-md5-hmac	带 MD5（HMAC 变量）的 ESP 验证算法
ah-sha-hmac	带 SHA（HMAC 变量）的 AH 验证算法	esp-3des	采用 3DES 的 ESP 加密算法	esp-sha-hmac	带 SHA（HMAC 变量）的 ESP 验证算法

IPSec 协议：ESP 和 AH。

ESP 和 AH 协议都为 IPSec 提供了安全服务。

ESP 提供了分组加密，以及可选的数据验证和抗重播服务。

AH 提供了数据验证和抗重播服务。

ESP 使用一个 ESP 头和一个 ESP 尾对受保护数据——或是一个完整的 IP 自寻址数据包（或仅是有效负载）——进行封装。AH 是嵌入在受保护数据中的；它将一个 AH 头直接插入在外部 IP 头后、内部 IP 数据包或有效负载前。隧道模式中要对整个 IP 数据报文进行封装和保护，而传送模式中只对 IP 数据报文中的有效负载进行封装/保护。要进一步了解这两种模式，请参阅 mode 命令的描述。

1）选择适当的变换。

IPSec 变换比较复杂。下面的提示能够帮助你选择适合自己情况的变换：

如果想要提供数据机密性，那么可以使用 ESP 加密变换。

如果想要提供对外部 IP 报头以及数据的数据验证，那么可以使用 AH 变换。

如果使用一个 ESP 加密变换，那么可以考虑使用 ESP 验证变换或 AH 变换来提供变换集合的验证服务。

如果想要数据验证功能（或使用 ESP 或使用 AH），可以选择 MD5 或 SHA 验证算法。SHA 算法比 MD5 要更安全，但速度更慢。

2）加密变换配置状态。

在执行了 crypto ipsec transform-set 命令以后，就将进入加密变换配置状态。在这种状态下，可以将模式改变到隧道模式或传输模式（这是可选的改变）。在做完这些改变以后，输入 exit 来返回到全局配置状态下。要深入了解这些可选改变的信息，请参看 mode 命令的详细阐述。

3）改变现存的变换。

如果在 transform-type 命令中为一个变换集合指定一个或多个变换，那么指定的这些变换将会替换掉变换集合中现存的变换。如果改变了 transform-type，改变将只被运用到引用了此变换集合的加密映射表上。但改变将不会被运用到现存的安全联盟上，会被用于新建立的安全联盟。如果想让新的设置立即生效，可以使用 clear crypto sa 命令来清除安全联盟数据库的部分或全部。

示例

以下例子定义了一个变换集合。

```
crypto ipsec transform-set one
transform-type esp-des esp-sha-hmac
```

实训 5　L2TP/PPTP VPN 的配置

实训目标

小宋的表哥在一家网吧负责网管工作，最近的一次聚会中，表哥提到网管的工作主要是维护网吧内网络设备及 PC 的正常运行。但经常出现他已经回到家中，网吧反馈有问题的情况。一般碰到这种情况，表哥就要赶回单位进行网络排查和调试。表哥想和小宋咨询一下有没有什么方法可以更加简便地针对远程网络进行维护。

小宋：表哥，这种情况对于网络维护人员或公司内有移动办公的用户很常见，一般通过 L2TP/PPTP VPN 就可以远程登录到内网进行维护管理了。因为 L2TP/PPTP 是 Windows 自带的标准 VPN 客户端协议，只要是 Windows 系统的 PC 都可以支持这两种 VPN 协议。

于是小宋就向表哥演示了 L2TP/PPTP VPN 的配置过程。

实训设备

1）DCR2655/2659 路由器 2 台。
2）CR-V35FC 1/ CR-V35MT 1 各 1 条，交叉双绞线 2 条。
3）PC 2 台。

实训拓扑

实训拓扑图如图 8-10 所示。

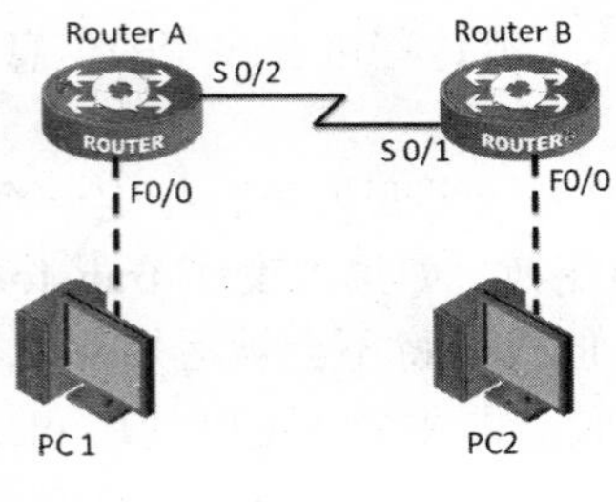

图　8-10

实训任务

任务 1：参照相关实训和表 8-12，配置所有接口的地址，并测试连通性。

表　8-12

主　机　名	F0/0	S0/1(DTE)	S0/2(DCE)	PC　地　址	默 认 网 关
Router A	192.168.0.1/24		192.168.1.1/24		
Router B	192.168.2.1/24	192.168.1.2/24			
PC1				192.168.0.2/24	192.168.0.1
PC2				192.168.2.2/24	192.168.2.1

任务 2：路由器 A 的配置。代码如下。

```
Router A#conf
```

```
Router A_config#int virtual-tunnel 0
Router A_config_vn0#ip address 172.16.1.2 255.255.255.0
Router A_config_vn0#ppp chap host test@dcn.net
Router A_config_vn0#ppp chap password 1234
Router A_config_vn0#exit
Router A_config#vpdn enable
Router A_config#vpdn-group 0
Router A_config_vpdn#request-dialin
Router A_config_vpdn#protocol l2tp
Router A_config_vpdn#initiate-to ip 192.168.1.2 priority 1
Router A_config_vpdn#domain dcn.net
Router A_config_vn0#exit
Router A_config#ip route 192.168.2.0 255.255.255.0 virtual-tunnel 0
```

任务 3：查看路由器 A 的配置。代码如下。

```
Router A#sh l2tp tunnel
L2TP Tunnel Information:
No active tunnels

Router A#sh l2tp session
L2TP Session Information:
No active sessions

Router A#sh int virtual-tunnel 0
Virtual-tunnel0 is up, line protocol is down
  Hardware is Unknown device
  MTU 1500 bytes, BW 100000 kbit, DLY 10000 usec
  Interface address is 172.16.1.2/24
  Encapsulation PPP, loopback not set
  Keepalive set(10 sec)
  LCP    Starting configuration exchange
  IPCP Listening -- waiting for remote host to attempt open
        local IP address: 172.16.1.2    remote IP address: 0.0.0.0
```

任务 4：路由器 B 的配置。代码如下。

```
Router B#conf
Router B_config#user test@dcn.net password 0 1234
Router B_config#ip local pool l2tppool 172.16.1.10 10
Router B_config#int virtual-template 0
Router B_config_vt0#ip address 172.16.1.1 255.255.255.0
```

```
Router B_config_vt0#ppp authen chap
Router B_config_vt0#peer default ip address pool l2tppool
Router B_config_vt0#exit
Router B_config#vpdn enable
Router B_config#vpdn-group 0
Router B_config_vpdn#accept-dialin
Router B_config_vpdn#protocol l2tp
Router B_config_vpdn#lcp-renegotiation
Router B_config_vpdn#port virtual-template 0
Router B_config_vpdn#exit
Router B_config#ip route 192.168.0.0 255.255.255.0 172.16.1.2
Router B_config#^Z
```

任务 5：查看路由器 B 的配置。代码如下。

```
Router B# sh run
Building configuration...

Current configuration:
!
!version 1.3.2E
service timestamps log date
service timestamps debug date
no service password-encryption
!
hostname Router-B
!
ip host a 192.168.1.1
ip host c 192.168.2.2
!
!
!
!
ip local pool l2tppool 172.16.1.10 10
!
aaa authentication ppp test local
username test@dcn.net password 0 1234
!
!
!
interface Virtual-template0
```

```
 ip address 172.16.1.1 255.255.255.0
 no ip directed-broadcast
 ppp authentication chap test
 peer default ip address pool l2tppool
!
interface FastEthernet0/0
 ip address 192.168.2.1 255.255.255.0
 no ip directed-broadcast
!
interface Serial1/0
 ip address 192.168.1.2 255.255.255.0
 no ip directed-broadcast
!
interface Async0/0
 no ip address
 no ip directed-broadcast
!
!
!
!
ip route 192.168.0.0 255.255.255.0 171.16.1.2
!
!
!
!
!
!
!
!
vpdn enable
!
vpdn-group 0
 accept-dialin
 lcp-renegotiation
 protocol l2tp
 local-name Digitalchina
 virtual-template 0
```

任务 6：测试。如图 8-11 所示。

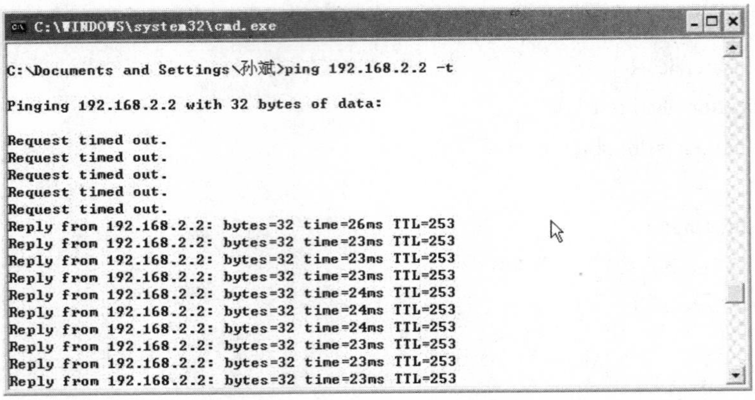

```
C:\WINDOWS\system32\cmd.exe

C:\Documents and Settings\孙斌>ping 192.168.2.2 -t

Pinging 192.168.2.2 with 32 bytes of data:

Request timed out.
Request timed out.
Request timed out.
Request timed out.
Request timed out.
Reply from 192.168.2.2: bytes=32 time=26ms TTL=253
Reply from 192.168.2.2: bytes=32 time=23ms TTL=253
Reply from 192.168.2.2: bytes=32 time=23ms TTL=253
Reply from 192.168.2.2: bytes=32 time=23ms TTL=253
Reply from 192.168.2.2: bytes=32 time=24ms TTL=253
Reply from 192.168.2.2: bytes=32 time=24ms TTL=253
Reply from 192.168.2.2: bytes=32 time=24ms TTL=253
Reply from 192.168.2.2: bytes=32 time=23ms TTL=253
Reply from 192.168.2.2: bytes=32 time=23ms TTL=253
Reply from 192.168.2.2: bytes=32 time=23ms TTL=253
```

图 8-11

注意事项和排错

1）注意两端参数要一致。

2）ACL 的作用是确定哪些数据需要经过 VPN。

3）密钥要交叉对应。

第 9 章　路由器综合应用进阶

实训 1　综合实验 1

实训目标

根据网络拓扑：

1）路由器背对背线缆两端 IP 地址分别为 1.1.1.1/24 和 1.1.1.2/24。PPP 封装，无验证。

2)路由器 R1 与交换机 S1 连接使用 f0/0 端口，交换机 S1 的 E0/0/23 口与路由器 R1F0/0 均为 trunk 端口。

3）交换机 S1 与 S2 之间链路也为 trunk 链路。

4）Vlan 10 成员 IP 地址范围 192.168.1.0/24，其中 R1 虚拟 Vlan 10 接口地址为 192.168.1.1/24。Vlan 20 成员 IP 地址范围 192.168.2.0/24，R1 虚拟 Vlan 20 接口地址为 192.168.2.1/24。

5）R2 使用 F0/0 与 PC5 连接，PC5 地址范围为 192.168.3.30/24，路由器接口地址 192.168.3.1/24。

6）实训中保证各 Vlan 中的 PC 与 192.168.3.30 都可以直接连通。

7）路由器之间使用 RIP 的直连引入方式进行设置。

实训设备

1）DCR2655/2659 路由器 2 台。

2）CR-V35FC 1/ CR-V35MT 1 各 1 条，交叉双绞线 2 条，直通线 5 条。

3）DCS3950 交换机 2 台。

4）PC 5 台。

实训拓扑

实训拓扑图如图 9-1 所示。

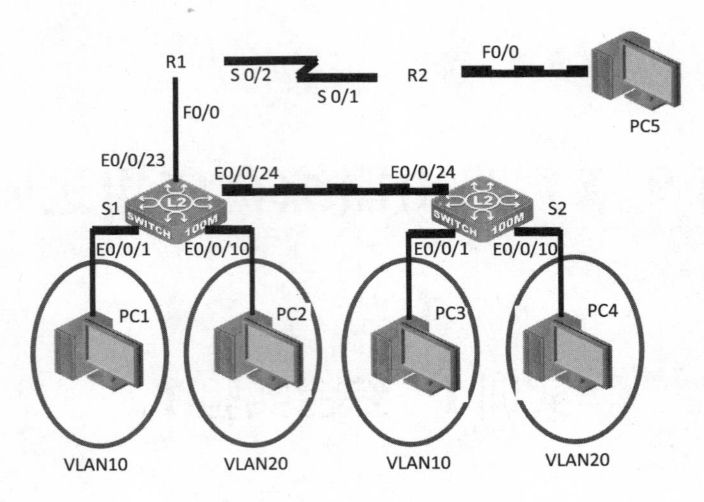

图 9-1

实训任务

任务 1：设置两台交换机 VLAN 并对封装端口进行配置。

S1：

```
switch#config
switch(Config)#hostname S1
S1(Config)#vlan 10
S1 (Config-Vlan10)#switchport interface ethernet 0/0/1-9
Set the port Ethernet0/0/1 access vlan 10 successfully
Set the port Ethernet0/0/2 access vlan 10 successfully
Set the port Ethernet0/0/3 access vlan 10 successfully
Set the port Ethernet0/0/4 access vlan 10 successfully
Set the port Ethernet0/0/5 access vlan 10 successfully
Set the port Ethernet0/0/6 access vlan 10 successfully
Set the port Ethernet0/0/7 access vlan 10 successfully
Set the port Ethernet0/0/8 access vlan 10 successfully
Set the port Ethernet0/0/9 access vlan 10 successfully
S1 (Config-Vlan10)#exit
S1 (Config)#vlan 20
S1 (Config-Vlan20)#switchport interface ethernet 0/0/10-20
Set the port Ethernet0/0/10 access vlan 20 successfully
Set the port Ethernet0/0/11access vlan 20 successfully
Set the port Ethernet0/0/12access vlan 20 successfully
```

```
Set the port Ethernet0/0/13 access vlan 20 successfully
Set the port Ethernet0/0/14access vlan 20 successfully
Set the port Ethernet0/0/15 access vlan 20 successfully
Set the port Ethernet0/0/16access vlan 20 successfully
Set the port Ethernet0/0/17 access vlan 20 successfully
Set the port Ethernet0/0/18 access vlan 20 successfully
Set the port Ethernet0/0/19access vlan 20 successfully
Set the port Ethernet0/0/20 access vlan 20 successfully
S1 (Config-Vlan20)#exit
S1 (Config)#interface ethernet 0/0/23
S1 (Config-Ethernet0/0/23)#switchport mode trunk
Set the port Ethernet0/0/23 mode TRUNK successfully
S1 (Config-Ethernet0/0/23)#switchport trunk native vlan 10
Set the port Ethernet0/0/23 native vlan 10 successfully
S1 (Config-Ethernet0/0/23)#switchport trunk allowed vlan all
set the port Ethernet0/0/23 allowed vlan successfully
S1 (Config-Ethernet0/0/23)#exit
S1 (Config)#interface ethernet 0/0/24
S1 (Config-Ethernet0/0/24)#switchport mode trunk
Set the port Ethernet0/0/24 mode TRUNK successfully
S1 (Config-Ethernet0/0/24)#switchport trunk native vlan 10
Set the port Ethernet0/0/24 native vlan 10 successfully
S1 (Config-Ethernet0/0/24)#switchport trunk allowed vlan all
set the port Ethernet0/0/24 allowed vlan successfully
S1 (Config-Ethernet0/0/24)#
```

S2：

```
switch(Config)#hostname S2
S2(Config)#no int vlan 1
S2 (Config)#vlan 10
S2 (Config-Vlan10)#switchport interface ethernet 0/0/1-9
Set the port Ethernet0/0/1 access vlan 10 successfully
Set the port Ethernet0/0/2 access vlan 10 successfully
Set the port Ethernet0/0/3 access vlan 10 successfully
Set the port Ethernet0/0/4 access vlan 10 successfully
Set the port Ethernet0/0/5 access vlan 10 successfully
Set the port Ethernet0/0/6 access vlan 10 successfully
Set the port Ethernet0/0/7 access vlan 10 successfully
Set the port Ethernet0/0/8 access vlan 10 successfully
```

```
Set the port Ethernet0/0/9 access vlan 10 successfully
S2 (Config-Vlan10)#exit
S2 (Config)#vlan 20
S2 (Config-Vlan20)#switchport interface ethernet 0/0/10-20
Set the port Ethernet0/0/10 access vlan 20 successfully
Set the port Ethernet0/0/11access vlan 20 successfully
Set the port Ethernet0/0/12access vlan 20 successfully
Set the port Ethernet0/0/13 access vlan 20 successfully
Set the port Ethernet0/0/14access vlan 20 successfully
Set the port Ethernet0/0/15 access vlan 20 successfully
Set the port Ethernet0/0/16access vlan 20 successfully
Set the port Ethernet0/0/17 access vlan 20 successfully
Set the port Ethernet0/0/18 access vlan 20 successfully
Set the port Ethernet0/0/19access vlan 20 successfully
Set the port Ethernet0/0/20 access vlan 20 successfully
S2 (Config-Vlan20)#exit
S2 (Config)#interface ethernet 0/0/24
S2 (Config-Ethernet0/0/24)#switchport mode trunk
Set the port Ethernet0/0/24 mode TRUNK successfully
S2 (Config-Ethernet0/0/24)#switchport trunk native vlan 10
Set the port Ethernet0/0/24 native vlan 10 successfully
S2 (Config-Ethernet0/0/24)#switchport trunk allowed vlan all
set the port Ethernet0/0/24 allowed vlan successfully
S2 (Config-Ethernet0/0/24)#exit
S2 (Config)#
```

验证配置：

S1：

```
S1#show vlan
VLAN Name            Type        Media      Ports
---- ----------- ---------- --------- ----------------------------------------
1    default         Static      ENET        Ethernet0/0/21          Ethernet0/0/22
                                             Ethernet0/0/23          Ethernet0/0/24(T)
                                             Ethernet0/0/25          Ethernet0/0/26
                                             Ethernet0/0/27          Ethernet0/0/28
10   VLAN0010        Static      ENET        Ethernet0/0/1           Ethernet0/0/2
                                             Ethernet0/0/3           Ethernet0/0/4
                                             Ethernet0/0/5           Ethernet0/0/6
```

```
                                              Ethernet0/0/7          Ethernet0/0/8
                                              Ethernet0/0/9          Ethernet0/0/23(T)
                                              Ethernet0/0/24
20    VLAN0020      Static       ENET         Ethernet0/0/10         Ethernet0/0/11
                                              Ethernet0/0/12         Ethernet0/0/13
                                              Ethernet0/0/14         Ethernet0/0/15
                                              Ethernet0/0/16         Ethernet0/0/17
                                              Ethernet0/0/18         Ethernet0/0/19
                                              Ethernet0/0/20         Ethernet0/0/23(T)
                                              Ethernet0/0/24(T)
```

S2：

```
S2#show vlan
VLAN Name               Type          Media        Ports
---- ------------ ---------- --------- ----------------------------------------
1     default      Static      ENET        Ethernet0/0/21         Ethernet0/0/22
                                           Ethernet0/0/23         Ethernet0/0/24(T)
                                           Ethernet0/0/25         Ethernet0/0/26
                                           Ethernet0/0/27         Ethernet0/0/28
10    VLAN0010      Static       ENET      Ethernet0/0/1          Ethernet0/0/2
                                           Ethernet0/0/3          Ethernet0/0/4
                                           Ethernet0/0/5          Ethernet0/0/6
                                           Ethernet0/0/7          Ethernet0/0/8
                                           Ethernet0/0/9          Ethernet0/0/24
20    VLAN0020      Static       ENET      Ethernet0/0/10         Ethernet0/0/11
                                           Ethernet0/0/12         Ethernet0/0/13
                                           Ethernet0/0/14         Ethernet0/0/15
                                           Ethernet0/0/16         Ethernet0/0/17
                                           Ethernet0/0/18         Ethernet0/0/19
                                           Ethernet0/0/20         Ethernet0/0/24(T)
```

任务 2：设置路由器以太网端口与 VLAN 成员可互通。代码如下。

```
Router#conf
Router_config#hostname R1
R1_config#interface f0/0.1      //进入路由器子接口
R1_config_f0/0.1#ip address 192.168.1.1 255.255.255.0      //为子接口添加 IP
R1_config_f0/0.1#encapsulation dot1Q 10      //为 VLAN10 的数据包封装 802.1Q 协议
R1_config_f0/0.1#exit
R1_config#interface f0/0.2      //进入路由器子接口
```

```
R1_config_f0/0.2#ip address 192.168.2.1 255.255.255.0     //为子接口添加 IP
R1_config_f0/0.2#encapsulation dot1Q 20   //为 VLAN20 的数据包封装 802.1Q 协议
R1_config_f0/0.2#exit
R1_config#
R1r#wr
Saving current configuration...
OK!
R1#
```

验证互通：

分别将 PC 192.168.1.10 接入到左侧交换机的 E0/0/1 和右侧交换机的 E0/0/1，验证与 192.168.1.1 的连通性。

分别将 PC 192.168.2.10 接入到左侧交换机的 E0/0/10 和右侧交换机的 E0/0/10，验证与 192.168.2.1 的连通性。

任务 3：配置路由器背对背链路，保证链路通畅。

左侧路由器：

```
R1#config
R1_config#interface serial 0/2
R1_config_s2/0#encapsulation ppp
R1_config_s2/0#physical-layer speed 64000
R1_config_s2/0#ip address 1.1.1.1 255.255.255.0
R1_config_s2/0#exit
R1_config#
```

右侧路由器：

```
Router#config
Router#hostname R2
R2_config#interface serial 0/1
R2_config_s1/0#encapsulation ppp
R2_config_s1/0#ip address 1.1.1.2 255.255.255.0
R2_config_s1/0#exit
R2_config#exit
R2#
```

验证连通性：

```
R2#ping 1.1.1.1
PING 1.1.1.1 (1.1.1.1): 56 data bytes
!!!!!
--- 1.1.1.1 ping statistics ---
5 packets transmitted, 5 packets received, 0% packet loss
round-trip min/avg/max = 20/22/30 ms
```

```
R2#
```

任务 4：配置右侧路由器以太网端口与 PC 192.168.3.30 可连通。代码如下。

```
R2#config
R2_config#interface fastethernet 0/0
R2_config_f0/0#ip address 192.168.3.1 255.255.255.0
R2_config_f0/0#exit
R2_config#exit
R2#2004-1-1 00:16:42 Configured from console 0 by DEFAULT
R2#
```

验证：

```
R2#ping 192.168.3.30
PING 192.168.3.30 (192.168.3.30): 56 data bytes
!!!!!
--- 192.168.3.30 ping statistics ---
5 packets transmitted, 5 packets received, 0% packet loss
round-trip min/avg/max = 0/0/0 ms
R2#
```

任务 5：配置路由器的 RIP，验证连通结果。

左侧路由器（同右侧路由器）：

```
R1#config
R1_config#router rip
R1_config_rip#network 1.1.1.0
R1_config_rip#redistribute connect
R1_config_rip#exit
R1_config#exit
```

验证路由表

左侧路由器：

```
R1#sh ip router
Codes: C - connected, S - static, R - RIP, B - BGP, BC - BGP connected
        D - DEIGRP, DEX - external DEIGRP, O - OSPF, OIA - OSPF inter area
        ON1 - OSPF NSSA external type 1, ON2 - OSPF NSSA external type 2
        OE1 - OSPF external type 1, OE2 - OSPF external type 2
        DHCP - DHCP type


VRF ID: 0
C       1.1.1.0/24          is directly connected, Serial0/2
R       1.1.1.1/32          [120,1] via 1.1.1.2(on Serial0/2)
C       1.1.1.2/32          is directly connected, Serial0/2
C       192.168.1.0/24      is directly connected, FastEthernet0/0.1
```

```
C        192.168.2.0/24          is directly connected, FastEthernet0/0.2
R1#
```

右侧路由器：

```
R2#sh ip router
Codes: C - connected, S - static, R - RIP, B - BGP, BC - BGP connected
       D - DEIGRP, DEX - external DEIGRP, O - OSPF, OIA - OSPF inter area
       ON1 - OSPF NSSA external type 1, ON2 - OSPF NSSA external type 2
       OE1 - OSPF external type 1, OE2 - OSPF external type 2
       DHCP - DHCP type

VRF ID: 0
C        1.1.1.0/24              is directly connected, Serial0/1
C        1.1.1.1/32              is directly connected, Serial0/1
R        1.1.1.2/32              [120,1] via 1.1.1.1(on Serial0/1)
R        192.168.1.0/24          [120,1] via 1.1.1.1(on Serial0/1)
R        192.168.2.0/24          [120,1] via 1.1.1.1(on Serial0/1)
C        192.168.3.0/24          is directly connected, FastEthernet0/0
R2#
```

验证 PC 的互通性。

略。

实训 2　综合实验 2

实训目标

根据网络拓扑：

1）PC1 与 PC2 能够互相 ping 通对方。

2）R1、R2、R3 之间使用 RIP 形成路由表。

3）PC1 与 PC2 之间的 FTP 数据使用 VPN 隧道通信，其他数据不必加密。

4）链路 R1-R2-R3 和 R1-R3 互为备份路由。

实训设备

1）DCR2655/2659 路由器 3 台。

2）CR-V35FC 1/ CR-V35MT 1 各 1 条，交叉双绞线 4 条。

3）PC 2 台。

实训拓扑

实训拓扑图如图 9-2 所示。

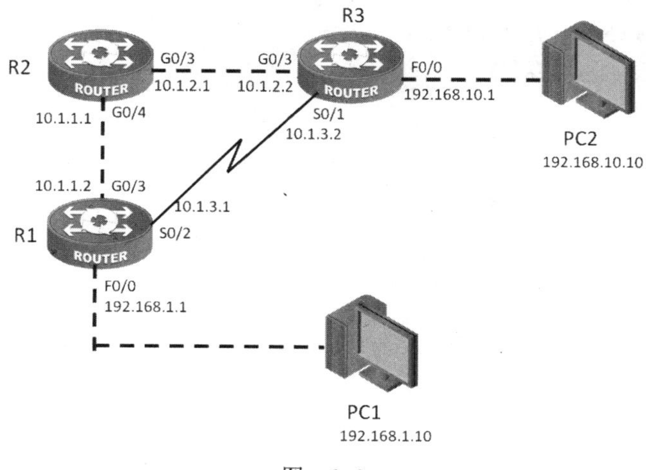

图　9-2

实训任务

任务 1：设置两台 PC 的地址。

任务 2：配置路由器的接口地址。

R1:

```
Router#conf
Router#hostname R1
R1_config#interface f0/0
R1_config_f0/0#ip address 192.168.1.1 255.255.255.0
R1_config_f0/0#no shutdown
R1_config_f0/0#exit
R1_config#interface g0/3
R1_config_g0/3#ip address 10.1.1.2 255.255.255.0
R1_config_g0/3#no shutdown
R1_config_g0/3#exit
R1_config#interface s0/2
R1_config_s0/2#ip address 10.1.3.1 255.255.255.0
R1_config_s0/2#physical-layer speed 64000
R1_config_s0/2#no shutdown
R1_config_s0/2#exit
```

```
R1_config#
```

R2:

```
Router#conf
Router_config#hostname R2
R2_config#interface g0/3
R2_config_g0/3#ip address 10.1.2.1 255.255.255.0
R2_config_g0/3#no shutdown
R2_config_g0/3#exit
R2_config#interface g0/4
R2_config_g0/4#ip address 10.1.1.1 255.255.255.0
R2_config_g0/4#no shutdown
R2_config_g0/4#exit
```

R3:

```
Router#conf
Router_config#hostname R3
R3_config#interface f0/0
R3_config_f0/0#ip address 192.168.10.1 255.255.255.0
R3_config_f0/0#no shutdown
R3_config_f0/0#exit
R3_config#interface g0/3
R3_config_g0/3#ip address 10.1.2.2 255.255.255.0
R3_config_g0/3#no shutdown
R3_config_g0/3#exit
R3_config#interface s0/1
R3_config_s0/1#ip address 10.1.3.2 255.255.255.0
R3_config_s0/1#no shutdown
R3_config_s0/1#exit
R3_config#
```

任务 3：查看路由表。

R1:

```
R1#show ip router
Codes: C - connected, S - static, R - RIP, B - BGP, BC - BGP connected
       D - BEIGRP, DEX - external BEIGRP, O - OSPF, OIA - OSPF inter area
       ON1 - OSPF NSSA external type 1, ON2 - OSPF NSSA external type 2
       OE1 - OSPF external type 1, OE2 - OSPF external type 2
       DHCP - DHCP type, L1 - IS-IS level-1, L2 - IS-IS level-2
VRF ID: 0
C     10.1.1.0/24     is directly connected, GigaEthernet0/3
C     10.1.3.0/24     is directly connected, Serial0/2
```

```
C          192.168.1.0/24          is directly connected, FastEthernet0/0
```

R2:

```
R2#
R2#show ip router
Codes: C - connected, S - static, R - RIP, B - BGP, BC - BGP connected
       D - BEIGRP, DEX - external BEIGRP, O - OSPF, OIA - OSPF inter area
       ON1 - OSPF NSSA external type 1, ON2 - OSPF NSSA external type 2
       OE1 - OSPF external type 1, OE2 - OSPF external type 2
       DHCP - DHCP type, L1 - IS-IS level-1, L2 - IS-IS level-2
VRF ID: 0
C          10.1.1.0/24          is directly connected, GigaEthernet0/4
C          10.1.2.0/24          is directly connected, GigaEthernet0/3
```

R3:

```
R3#
R3#show ip router
Codes: C - connected, S - static, R - RIP, B - BGP, BC - BGP connected
       D - BEIGRP, DEX - external BEIGRP, O - OSPF, OIA - OSPF inter area
       ON1 - OSPF NSSA external type 1, ON2 - OSPF NSSA external type 2
       OE1 - OSPF external type 1, OE2 - OSPF external type 2
       DHCP - DHCP type, L1 - IS-IS level-1, L2 - IS-IS level-2
VRF ID: 0
C          10.1.2.0/24          is directly connected, GigaEthernet0/3
C          10.1.3.0/24          is directly connected, Serial0/1
C          192.168.10.0/24      is directly connected, FastEthernet0/0
R3#
```

任务 4：配置路由器 RIP。

R1:

```
R1#conf
R1_config#router rip
R1_config_rip#network 192.168.1.0
R1_config_rip#network 10.1.1.0
R1_config_rip#network 10.1.3.0
R1_config_rip#exit
R1_config#
```

R2：

```
R2#conf
R2_config#router rip
R2_config_rip#network 10.1.1.0
R2_config_rip#network 10.1.2.0
```

```
R2_config_rip#exit
R2_config#
```

R3:

```
R3#conf
R3_config#router rip
R3_config_rip#network 192.168.10.0
R3_config_rip#network 10.1.2.0
R3_config_rip#network 10.1.3.0
R3_config_rip#exit
R3_config#
```

任务 5：查看路由表。

R1:

```
R1#show ip router
Codes: C - connected, S - static, R - RIP, B - BGP, BC - BGP connected
        D - BEIGRP, DEX - external BEIGRP, O - OSPF, OIA - OSPF inter area
        ON1 - OSPF NSSA external type 1, ON2 - OSPF NSSA external type 2
        OE1 - OSPF external type 1, OE2 - OSPF external type 2
        DHCP - DHCP type, L1 - IS-IS level-1, L2 - IS-IS level-2
VRF ID: 0
C       10.1.1.0/24        is directly connected, GigaEthernet0/3
R       10.1.2.0/24        [120,1] via 10.1.3.2(on Serial0/2)
                             [120,1] via 10.1.1.1(on GigaEthernet0/3)
C       10.1.3.0/24        is directly connected, Serial0/2
C       192.168.1.0/24     is directly connected, FastEthernet0/0
R       192.168.10.0/24    [120,1] via 10.1.3.2(on Serial0/2)
```

R2:

```
R2#show ip router
Codes: C - connected, S - static, R - RIP, B - BGP, BC - BGP connected
        D - BEIGRP, DEX - external BEIGRP, O - OSPF, OIA - OSPF inter area
        ON1 - OSPF NSSA external type 1, ON2 - OSPF NSSA external type 2
        OE1 - OSPF external type 1, OE2 - OSPF external type 2
        DHCP - DHCP type, L1 - IS-IS level-1, L2 - IS-IS level-2
VRF ID: 0
C       10.1.1.0/24        is directly connected, GigaEthernet0/4
C       10.1.2.0/24        is directly connected, GigaEthernet0/3
R       10.1.3.0/24        [120,1] via 10.1.2.2(on GigaEthernet0/3)
                             [120,1] via 10.1.1.2(on GigaEthernet0/4)
R       192.168.1.0/24     [120,1] via 10.1.1.2(on GigaEthernet0/4)
R       192.168.10.0/24    [120,1] via 10.1.2.2(on GigaEthernet0/3)
```

R3:

```
R3#show ip router
Codes: C - connected, S - static, R - RIP, B - BGP, BC - BGP connected
       D - BEIGRP, DEX - external BEIGRP, O - OSPF, OIA - OSPF inter area
       ON1 - OSPF NSSA external type 1, ON2 - OSPF NSSA external type 2
       OE1 - OSPF external type 1, OE2 - OSPF external type 2
       DHCP - DHCP type, L1 - IS-IS level-1, L2 - IS-IS level-2
VRF ID: 0
R      10.1.1.0/24         [120,1] via 10.1.3.1(on Serial0/1)
                             [120,1] via 10.1.2.1(on GigaEthernet0/3)
C      10.1.2.0/24         is directly connected, GigaEthernet0/3
C      10.1.3.0/24         is directly connected, Serial0/1
R      192.168.1.0/24      [120,1] via 10.1.3.1(on Serial0/1)
C      192.168.10.0/24     is directly connected, FastEthernet0/0
```

任务 6：检验 PC 连通性。代码如下。

```
C:\Documents and Settings\Administrator>ping 192.168.10.10
Pinging 192.168.10.10 with 32 bytes of data:
Reply from 192.168.10.10: bytes=32 time=21ms TTL=62
Reply from 192.168.10.10: bytes=32 time=20ms TTL=62
Reply from 192.168.10.10: bytes=32 time=21ms TTL=62
Reply from 192.168.10.10: bytes=32 time=21ms TTL=62
Ping statistics for 192.168.10.10:
    Packets: Sent = 4, Received = 4, Lost = 0 (0% loss),
Approximate round trip times in milli-seconds:
    Minimum = 20ms, Maximum = 21ms, Average = 20ms
C:\Documents and Settings\Administrator>
```

任务 7：配置并查看访问列表。

R1:

```
R1#conf
R1_config#ip access-list extended for_vpn
R1_config_ext_nacl#permit tcp  192.168.1.10  255.255.255.255  192.168.10.10  255.255.255.255 eq 21
R1_config_ext_nacl#deny ip any any

R1#show ip access-lists
Extended IP access list for_vpn
 permit tcp 192.168.1.10 255.255.255.255 192.168.10.10 255.255.255.255 eq ftp
 deny    ip any any
R1#
```

R3:

```
R3#conf
R3_config#ip access-list extended for_vpn
R3_config_ext_nacl#permit  tcp  192.168.10.10  255.255.255.255  192.168.1.10  255.255.255.255 eq 21
R3_config_ext_nacl#deny ip any any

R3#show ip access-lists
Extended IP access list for_vpn
 permit tcp 192.168.10.10 255.255.255.255 192.168.1.10 255.255.255.255 eq ftp
 deny     ip any any
R3#
```

任务 8：配置 VPN（通过串口）。

R1:

```
R1#conf
R1_config#
R1_config#crypt isakmp key dcnu 10.1.3.2 255.255.255.0
R1_config#crypto ipsec transform-set r1
R1_config_crypto_trans#mode tunnel
R1_config_crypto_trans#transform-type esp-des esp-md5-hmac
R1_config_crypto_trans#exit
R1_config#crypto map r1 10 ipsec-isakmp
R1_config_crypto_map#set transform-set r1
R1_config_crypto_map#set peer 10.1.3.2
R1_config_crypto_map#match address for_vpn
R1_config_crypto_map#exit
R1_config#interface s0/2
R1_config_s0/2#crypto map r1
R1_config_s0/2#^Z
```

R3：

```
R3#conf
R3_config#crypto isakmp key dcnu 10.1.3.1 255.255.255.0
R3_config#crypto ipsec transform-set r1
R3_config_crypto_trans#mode tunnel
R3_config_crypto_trans#transform-type esp-des esp-md5-hmac
R3_config_crypto_trans#exit
R3_config#crypto map r1 10 ipsec-isakmp
R3_config_crypto_map#set transform-set r1
R3_config_crypto_map#set peer 10.1.3.1
R3_config_crypto_map#match address for_vpn
R3_config_crypto_map#exit
```

```
R3_config#interface s0/1
R3_config_s0/1#crypto map r1
R3_config_s0/1#^Z
```

任务 9：配置 VPN 备份（通过 G0/3 口）。

R1:

```
R1#conf
R1_config#crypt isakmp key dcnu 10.1.2.2 255.255.255.0
R1_config#crypto ipsec transform-set r2
R1_config_crypto_trans#transform-type esp-des esp-md5-hmac
R1_config_crypto_trans#exit
R1_config#crypto map r2 10 ipsec-isakmp
R1_config_crypto_map#set transform-set r2
R1_config_crypto_map#set peer 10.1.2.2
R1_config_crypto_map#match address for_vpn
R1_config_crypto_map#exit
R1_config#interface g0/3
R1_config_g0/3#crypto map r2
R1_config_g0/3#exit
```

R3:

```
R3#conf
R3_config#crypt isakmp key dcnu 10.1.1.2 255.255.255.0
R3_config#crypto ipsec transform-set r2
R3_config_crypto_trans#transform-type esp-des esp-md5-hmac
R3_config_crypto_trans#exit
R3_config#crypto map r2 10 ipsec-isakmp
R3_config_crypto_map#set transform-set r2
R3_config_crypto_map#set peer 10.1.1.2
R3_config_crypto_map#match address for_vpn
R3_config_crypto_map#exit
R3_config#interface g0/3
R3_config_g0/3#crypto map r2
R3_config_g0/3#exit
```

任务 10：验证结果。

```
R1#show crypto ipsec sa
Interface: Serial0/2
Crypto map name:r1 ,  local addr. 10.1.3.1
local   ident (addr/mask/prot/port): (192.168.1.10/255.255.255.255/6/0)
 remote ident (addr/mask/prot/port): (192.168.10.10/255.255.255.255/6/21)
 local crypto endpt.: 10.1.3.1,  remote crypto endpt.: 10.1.3.2
```

```
Interface: GigaEthernet0/3
Crypto map name:r2 ,   local addr. 10.1.1.2
local    ident (addr/mask/prot/port): (192.168.1.10/255.255.255.255/6/0)
 remote ident (addr/mask/prot/port): (192.168.10.10/255.255.255.255/6/21)
 remote ident (addr/mask/prot/port): (192.168.10.10/255.255.255.255/6/21)

R3#show crypto ipsec sa
Interface: Serial0/1
Crypto map name:r1 ,   local addr. 10.1.3.2
local    ident (addr/mask/prot/port): (192.168.10.10/255.255.255.255/6/0)
 remote ident (addr/mask/prot/port): (192.168.1.10/255.255.255.255/6/21)
 local crypto endpt.: 10.1.3.2,   remote crypto endpt.: 10.1.3.1
Interface: GigaEthernet0/3
Crypto map name:r2 ,   local addr. 10.1.2.2
local    ident (addr/mask/prot/port): (192.168.10.10/255.255.255.255/6/0)
 remote ident (addr/mask/prot/port): (192.168.1.10/255.255.255.255/6/21)
 local crypto endpt.: 10.1.2.2,   remote crypto endpt.: 10.1.1.2
```

任务 11：PC 机互 PING。

```
C:\Documents and Settings\Administrator>ping 192.168.10.10
Pinging 192.168.10.10 with 32 bytes of data:
Reply from 192.168.10.10: bytes=32 time=12ms TTL=62
Reply from 192.168.10.10: bytes=32 time=12ms TTL=62
Reply from 192.168.10.10: bytes=32 time=11ms TTL=62
Reply from 192.168.10.10: bytes=32 time=11ms TTL=62

Ping statistics for 192.168.10.10:
    Packets: Sent = 4, Received = 4, Lost = 0 (0% loss),
Approximate round trip times in milli-seconds:
    Minimum = 11ms, Maximum = 12ms, Average = 11ms
此时通过串口实现，shut 掉串口,然后继续 PING:
C:\Documents and Settings\Administrator>ping 192.168.10.10 -t
Pinging 192.168.10.10 with 32 bytes of data:
Reply from 192.168.1.1: Destination host unreachable.
Reply from 192.168.1.1: Destination host unreachable.
Reply from 192.168.1.1: Destination host unreachable.
Reply from 192.168.1.1: Destination host unreachable.
Reply from 192.168.1.1: Destination host unreachable.
Reply from 192.168.10.10: bytes=32 time=2ms TTL=61
Reply from 192.168.10.10: bytes=32 time=1ms TTL=61
```

```
Reply from 192.168.10.10: bytes=32 time=1ms TTL=61
Reply from 192.168.10.10: bytes=32 time=1ms TTL=61
Reply from 192.168.10.10: bytes=32 time=1ms TTL=61
Reply from 192.168.10.10: bytes=32 time=1ms TTL=61

Ping statistics for 192.168.10.10:
    Packets: Sent = 22, Received = 22, Lost = 0 (0% loss),
Approximate round trip times in milli-seconds:
    Minimum = 1ms, Maximum = 2ms, Average = 0ms
```

参 考 文 献

[1] 张宏科．路由器原理与技术[M]．北京：高等教育出版社，2010．
[2] 颜谦和．交换机与路由技术[M]．北京：清华大学出版社，2011．